Lecture Notes
in Business Information Processing

278

More information about this series at http://www.springer.com/series/7911

Derek Bridge · Heiner Stuckenschmidt (Eds.)

E-Commerce and Web Technologies

17th International Conference, EC-Web 2016
Porto, Portugal, September 5–8, 2016
Revised Selected Papers

 Springer

Editors
Derek Bridge
The Insight Centre for Data Analytics
University College Cork
Cork
Ireland

Heiner Stuckenschmidt
Data and Web Science Group
University of Mannheim
Mannheim
Germany

ISSN 1865-1348   ISSN 1865-1356 (electronic)
Lecture Notes in Business Information Processing
ISBN 978-3-319-53675-0   ISBN 978-3-319-53676-7 (eBook)
DOI 10.1007/978-3-319-53676-7

Library of Congress Control Number: 2017931544

Printed on acid-free paper

This Springer imprint is published by Springer Nature
The registered company is Springer International Publishing AG
The registered company address is: Gewerbestrasse 11, 6330 Cham, Switzerland

Preface

EC-Web is an international scientific conference series devoted to technology-related aspects of e-commerce and e-business. The 17th edition of the conference, EC-Web 2016, took place in Porto, Portugal, in September 2016 and served as a forum to bring together researchers and practitioners to present and discuss recent advances in their fields. The conference series historically covers the following areas:

- Search, comparison, and recommender systems
- Preference representation and reasoning
- Semantic-based systems, ontologies, and linked data
- Agent-based systems, negotiation, and auctions
- Social Web and social media in e-commerce
- Computational advertising
- E-commerce infrastructures and cloud-based services
- Service modelling and engineering
- Business processes, Web services, and service-oriented architectures
- E-business architectures
- Emerging business models, software as a service, mobile services
- Security, privacy, and trust
- Case studies

This year, the conference program focused on five main topic: recommender systems, product data on the Web, business processes and Web services and cloud computing, and data analysis. The works presented at the conference reflect recent trends in different subfields related to e-commerce and Web technologies, which can be summarized as follows.

- In the Web era, recommender systems play a fundamental role in helping users deal with issues related to information overload. Recent trends have emphasized the role of the user in these systems, with ways of giving users more control over the recommendations they receive and helping them to visualize the output of the recommender system.
- Product data is one of the most obvious manifestations of doing e-commerce on the Web. Techniques for extracting product data from website content and turning the captured data into canonical representations will extend the reach of e-commerce technologies.
- Business processes embody huge amounts of valuable expertise, which needs to be captured and stored. There is then a need to support search through repositories of business processes so that the expertise can be reused. In a similar vein, there is a growing volume of Web and cloud services that can be composed into larger services. This requires their retrieval, with regard to their inter-operability and quality.

- Finally, user activity on the Web produces growing quantities of data such as clickstreams and purchase histories. We need techniques for efficiently and accurately mining new knowledge from this data for applications such as recommendation and fraud detection.

Out of the contributions of the conference, we selected 12 papers to be included in the proceedings. The volume also includes one invited paper from the conference's keynote speaker. The accepted papers are organized in three themes:

The section on "Recommender Systems" in the proceedings contains four papers. Jorge et al. provide a survey of recommender systems with a focus on implicit feedback and online stream-based approaches. Their paper includes a discussion of methods for forgetting earlier data in the stream and for evaluating recommenders in this setting. This paper is an invited paper and is based on the talk given at the conference by our keynote speaker, Alípio M. Jorge. We are grateful to Prof. Jorge for taking the time to contribute to the conference in this way. Among the other papers in the recommender system section, Jannach et al. review different ways in which recommender systems can give the user more control over the recommendation process, and present a user survey concerning the control mechanisms available in Amazon.com. Deldjoo et al. extract features from movie frames and combine these using canonical correlation analysis with more conventional textual data about the movies, to build a high-performing hybrid. Finally, Richthammer and Pernul describe their use of treemaps for presenting a set of recommendations: The two-dimensional presentation can convey more information visually than can a conventional recommmendation list.

The next sections is concerned with "Data Management and Data Analysis" on the Web. Horch et al. manually analyze 50 different e-commerce websites, using descriptive statistics to give insight into the way that these shops structure and distribute product data across their sites, for example, that there may be up to around 15 different kinds of prices. Petrovski et al. describe a corpus of product data they produced that can act as a gold standard in tasks such as product matching: The corpus contains around 500 products and, for 150 of them, 1,500 positive and 73,500 negative correspondences between the products. de Amorim et al. offer a new fast algorithm for mining frequent itemsets from clickstreams. The algorithm is evaluated in a news article recommender system. Jiao et al. also mine clickstreams along with add-to-basket and purchase events, to investigate the value of search re-ranking strategies. Lastly, Lima and Pereira investigate the interplay between resampling and feature selection in fraud detection tasks.

The papers in the "Business Processes, Web Services, and Cloud Computing" section of the proceedings investigate the retrieval of business processes and the composition of Web services. Ordoñez et al. present a new way of indexing business processes, inspired by n-grams and capturing both the linguistic and structural aspects of the business process. In the case of cloud services, Ferrarons et al. define interoperability metrics such as quantitative coverage metrics and qualitative quality metrics. da Silva et al. present an approach to service composition that uses evolutionary computation techniques that work on services represented as directed acyclic graphs.

Finally, Kasmi et al. survey work that uses recommender systems techniques for Web services composition. The survey covers dimensions such as context, social networks, time, and interactivity, among others.

We are grateful to all authors, especially the ones who presented the papers at the conference, and to the DEXA conference organizers and Springer.

January 2017   Derek Bridge
                                                      Heiner Stuckenschmidt

Organization

EC-Web 2016 was organized in the context of the 27th International Conference on Database and Expert Systems Applications (DEXA 2016) and took place during September 5–6, 2016, in Porto.

Conference Organization

Co-chairs

Derek Bridge   Insight Centre for Data Analytics,
                                  University College Cork, Ireland
Heiner Stuckenschmidt   University of Mannheim, Germany

Program Committee

Burke, Robin   DePaul University, USA
Chen, Li   Hong Kong Baptist University, SAR China
Cuzzocrea, Alfredo   University of Trieste, Italy
De Luca, Ernesto William   FH Potsdam, Germany
Felfernig, Alexander   Graz University of Technology, Austria
García, José María   University of Seville, Spain
Jannach, Dietmar   TU Dortmund, Germany
Jorge, Alípio Mário   University of Porto, Portugal
Meusel, Robert   University of Mannheim, Germany
Monica, Dragoicea   Politehnica University of Bucharest, Romania
O'Mahony, Michael   University College Dublin, Ireland
Parra, Denis   Pontificia Universidad Catolica de Chile, Chile
Paulheim, Heiko   University of Mannheim, Germany
Quijano Sanchez, Lara   Universidad Complutense de Madrid, Spain
Vidal, Maria Esther   Universidad Simon Bolivar, Venezuela
Zanker, Markus   Free University of Bozen-Bolzano, Italy

Contents

Recommender Systems

Scalable Online Top-N Recommender Systems. 3
  Alípio M. Jorge, João Vinagre, Marcos Domingues, João Gama,
  Carlos Soares, Pawel Matuszyk, and Myra Spiliopoulou

User Control in Recommender Systems: Overview and Interaction
Challenges. 21
  Dietmar Jannach, Sidra Naveed, and Michael Jugovac

How to Combine Visual Features with Tags to Improve Movie
Recommendation Accuracy?. 34
  Yashar Deldjoo, Mehdi Elahi, Paolo Cremonesi,
  Farshad Bakhshandegan Moghaddam, and Andrea Luigi Edoardo Caielli

Explorative Analysis of Recommendations Through Interactive
Visualization 46
  Christian Richthammer and Günther Pernul

Data Management and Data Analysis

An E-Shop Analysis with a Focus on Product Data Extraction 61
  Andrea Horch, Andreas Wohlfrom, and Anette Weisbecker

The WDC Gold Standards for Product Feature Extraction and Product
Matching. 73
  Petar Petrovski, Anna Primpeli, Robert Meusel, and Christian Bizer

MFI-TransSW+: Efficiently Mining Frequent Itemsets in Clickstreams. 87
  Franklin A. de Amorim, Bernardo Pereira Nunes, Giseli Rabello Lopes,
  and Marco A. Casanova

Reranking Strategies Based on Fine-Grained Business User Events
Benchmarked on a Large E-commerce Data Set 100
  Yang Jiao, Bruno Goutorbe, Matthieu Cornec, Jeremie Jakubowicz,
  Christelle Grauer, Sebastien Romano, and Maxime Danini

Feature Selection Approaches to Fraud Detection in e-Payment Systems 111
  Rafael Franca Lima and Adriano C.M. Pereira

Business Processes, Web Services and Cloud Computing

Multimodal Indexing and Search of Business Processes Based on
Cumulative and Continuous N-Grams   129
   Hugo Ordoñez, Armando Ordoñez, Carlos Cobos, and Luis Merchan

Scoring Cloud Services Through Digital Ecosystem Community Analysis . . .   142
   *Jaume Ferrarons, Smrati Gupta, Victor Muntés-Mulero,
Josep-Lluis Larriba-Pey, and Peter Matthews*

Handling Branched Web Service Composition with a QoS-Aware
Graph-Based Method.   154
   *Alexandre Sawczuk da Silva, Hui Ma, Mengjie Zhang,
and Sven Hartmann*

Recommendation in Interactive Web Services Composition:
A State-of-the-Art Survey.   170
   Meriem Kasmi, Yassine Jamoussi, and Henda Hajjami Ben Ghézala

Author Index   183

Recommender Systems

Scalable Online Top-N Recommender Systems

Alípio M. Jorge[1,2(✉)], João Vinagre[1], Marcos Domingues[3], João Gama[1,2],
Carlos Soares[1,2], Pawel Matuszyk[4], and Myra Spiliopoulou[4]

[1] LIAAD/CESE, INESC TEC, Porto, Portugal
[2] FCUP/FEP/FEUP, Universidade do Porto, Porto, Portugal
amjorge@fc.up.pt
[3] Department of Informatics, State University of Maringá, Maringá, Brazil
[4] Knowledge Management and Discovery,
Otto-von-Guericke University, Magdeburg, Germany

Abstract. Given the large volumes and dynamics of data that recommender systems currently have to deal with, we look at online stream based approaches that are able to cope with high throughput observations. In this paper we describe work on incremental neighborhood based and incremental matrix factorization approaches for binary ratings, starting with a general introduction, looking at various approaches and describing existing enhancements. We refer to recent work on forgetting techniques and multidimensional recommendation. We will also focus on adequate procedures for the evaluation of online recommender algorithms.

1 Introduction

"Recommender system" is a designation that is currently used in more than one sense. Nevertheless, it typically refers to an information system, or part of it, that assists users in making choices from a very large catalog of items [25]. Although quite common in e-commerce, recommendation capabilities appear in a great variety of applications and situations, including e-learning [15], health [8], human resource management [30] and public transports [19]. Central to each recommender systems is the filtering criterion that is applied to the whole collection of available items. This criterion can be defined in many different ways, depending on the approach and on the available information at the time of recommendation.

The impact of recommender systems is now clearly recognized by companies and society. Amazon, a company who had a major role in the popularization of modern automatic recommenders, acknowledges that 35% of consumer choices are driven by recommendations [20]. Netflix reports a considerably larger value of 75%, which is comprehensible given that the service is payed via a flat rate and the consumption of movies and tv series can be done in a more streaming fashion. It is also recognized that displayed recommendations can improve the reputation of items, even if the predicted rating is low [23]. Recommendations may reduce the variety of items that users consume (the filter bubble effect), but

© Springer International Publishing AG 2017
D. Bridge and H. Stuckenschmidt (Eds.): EC-Web 2016, LNBIP 278, pp. 3–20, 2017.
DOI: 10.1007/978-3-319-53676-7_1

may also allow the discovery of novel music and books, timidly moving towards the *serendipity* effect. Recommendations may also promote a *rich get richer* phenomenon and make marketeers worry about getting noticed by algorithms rather than getting noticed by users [13].

The two central entities of a recommender system are *users* and *items*. The general setting is that users interact with items, are interested in new items and items are recommended to users. The algorithms that produce the recommendations rely on models of both users and items. The modeling of users is called *profiling* and is based on the selection of descriptive measurable dimensions. Among these we may use demographic information, user activity and existing interaction with items (e.g. user bought item, user viewed item description, user listened to music), user's social connections, textual reviews, numerical ratings, etc. Items are also modeled on the basis of item content, including meta-data, item textual description and, if available, images or audio. Any item usage data (user-item interactions) can also be used to characterize and profile items. In general, an automatic recommendation can have the form of a predicted score (estimation of how much a user likes an item), a *Top-N* list with the N most likely preferred items or even a structured suggestion such as trip with multiple stops.

In this paper we describe our contributions to the problem of Top-N recommendation exploring interaction data. We focus on online/incremental algorithms, as well as techniques that help deal with temporal issues and the exploration of multiple dimensions.

2   Profiling

It is customary to divide the flavors of user and item profiling into *content based* and *usage based*. This results in *content based filtering* and *collaborative filtering*. An example of the former is: "If the user likes prize-winning novelists then recommend a book written by a Nobel prize of literature". In this case, characteristics of item content are used to produce the matching. In collaborative filtering, information is gathered from a set of users and their interaction with items. Since no content data has to be used, collaborative filtering recommenders do not have to possess any kind of capability of "understanding" the content and merely rely on the "wisdom of the crowd". Although this is, more often than not, an advantage, it can also be a source of the "tyranny of the average", making life harder to users in behavior niches. It also suffers more easily of difficulties in making recommendations to new users or in recommending new items, the so called *cold start* problems.

In collaborative filtering approaches, profiling a specific user can be reduced to collecting the interaction of that user with items. One possible interaction is the rating of items by the users. These *ratings* are values in a pre-defined range, e.g. one to five stars, and can be treated as continuous. In this case the profile of the user can be represented by a real valued vector with as many dimensions as items. Ratings express explicit opinions of users, require more effort from the user and are, for that reason, relatively hard to obtain. They provide, however, high quality information. Another possible interaction is an action of the user over an item that indicates some interest or preference of the user for that item. One common example of such an action is "user bought item". The profile of an user is then a *binary* vector with ones on the dimensions corresponding to preferred items. Another way of seeing the profile in this binary setting is as the set of items the user interacted with. The binary setting is also sometimes called *unary* since only positive information is collected. Binary data is typically much cheaper to obtain than ratings data since it can result from the normal interaction of users with items, as they browse the catalog or decide to buy. An intermediate possibility is to automatically infer the degree of preference of an user from the natural interactions with the items. For example, we can infer that a user likes a music track less or more given the number of times it is played. These are called *implicit ratings*.

3 Baseline Collaborative Filtering Algorithms

Most state-of-the-art Collaborative Filtering (CF) algorithms are based on either neighborhood methods or matrix factorization. Fundamentally, these differ on the strategy used to process user feedback. This user feedback can be conceptually seen as a user-item matrix, known in most literature as the *ratings matrix*, in which cells contain information about user preferences over items.

3.1 Neighborhood-Based Algorithms

Neighborhood-based algorithms essentially compute user or item neighborhoods using a user defined similarity measure. Typical measures include the cosine and Pearson correlation [27]. If the rows of the user-item matrix R represent users, and the columns correspond to items, similarity between two users u and v is obtained by measuring the similarity between the rows corresponding to those users, R_u and R_v. Similarity between two items i and j can be obtained between the columns corresponding to those items R_i and R_j. Recommendations are computed by searching and aggregating through the user's or item's k nearest neighbors. The optimal value of k is data dependent and can be obtained using cross-validation. The main advantages of neighborhood methods are their simplicity and ease of implementation, as well as the trivial explainability of recommendations – user and item similarities are intuitive concepts.

User-based CF exploits similarities between users to form user neighborhoods. For example, given two users u and v, the cosine similarity between them takes the rows of the ratings matrix R_u and R_v as vectors in a space with dimension equal to the number of items rated by both u and v:

$$\text{sim}(u,v) = \cos(R_u, R_v) = \frac{R_u \cdot R_v}{||R_u|| \times ||R_v||} = \frac{\sum_{i \in I_{uv}} R_{ui} R_{vi}}{\sqrt{\sum_{i \in I_u} R_{ui}^2}\sqrt{\sum_{i \in I_v} R_{vi}^2}} \tag{1}$$

where $R_u \cdot R_v$ represents the dot product between R_u and R_v, I_u and I_v are the sets of items rated by u and v respectively and $I_{uv} = I_u \cap I_v$ is the set of items rated by both users u and v. Other similarity measures can be used, such as the Pearson Correlation. Euclidean measures are typically not used, given the very high dimensionality of the problem.

To compute the rating prediction $\hat{R}_{ui}$ given by the user u to item i, an aggregating function is used that combines the ratings given to i by the subset $K_u \subseteq U$ of the k nearest neighbors of u:

$$\hat{R}_{ui} = \frac{\sum_{v \in K_u} \text{sim}(u, v) R_{vi}}{\sum_{v \in K_u} \text{sim}(u, v)} \tag{2}$$

$$\hat{R}_{ui} = \bar{R}_u + \frac{\sum_{v \in K_u} \text{sim}(u, v)(R_{vi} - \bar{R}_v)}{\sum_{v \in K_u} \text{sim}(u, v)} \tag{3}$$

Equation (2) performs a simple weighted average, in which weights are given by the similarities between u and v. Equation (3) incorporates the average ratings given by u and v in order to minimize differences in how users use the rating scale – i.e. some users tend give higher/lower ratings than others in average.

Similarity between items can also be explored to provide recommendations [14,27]. One practical way of looking at it is simply to transpose the user-item ratings matrix and then apply the exact same method. The result of the training algorithm will be an item-item similarity matrix.

3.1.1 Neighborhood-Based CF for Binary Data

The methods in Sect. 3.1 are designed to work with numerical ratings. Neighborhood-based CF for binary data can actually be seen as a special case of neighborhood-based CF for ratings, by simply considering $R_{ui} = 1$ for all observed (u, i) user-item pairs and $R_{ui} = 0$ for all other cells in the feedback matrix R. Both notation and implementation can be simplified with this. For example, the cosine for binary ratings can be written as:

$$\text{sim}(u, v) = \cos(R_u, R_v) = \frac{\sum_{i \in I_{uv}} R_{ui} R_{vi}}{\sqrt{\sum_{i \in I_u} R_{ui}^2} \sqrt{\sum_{i \in I_v} R_{vi}^2}} = \frac{|(I_u \cap I_v)|}{\sqrt{|I_u|} \sqrt{|I_v|}} \tag{4}$$

where I_u and I_v the set of items that are observed with u and j, respectively.

The cosine formulation in (4) reveals that it is possible to calculate user-user similarities using simple occurrence and co-occurrence counts. A user u is said to co-occur with user v for every item i they both occur with. Similarly, in the item-based case, an item i is said to co-occur with item j every time they both occur with a user u.

Rating predictions can be as well made using (2). The value of $\hat{R}_{ui}$ will be a score between 0 and 1, by which a list of candidate items for recommendation can be sorted in descending order for every user.

3.2   Matrix Factorization

Matrix Factorization recommendation algorithms are inspired by Latent Semantic Indexing (LSI) [3], an information retrieval technique to index large collections of text documents. LSI performs the Singular Value Decomposition (SVD) of large document-term matrices. In a recommendation problem, the same technique can be used in the user-item matrix, uncovering a latent feature space that is common to both users and items. The problem with SVD is that classic factorization algorithms, such as Lanczos methods, are not defined for sparse matrices, which raises problems on how to compute the factorization.

As an alternative to classic SVD, optimization methods have been proposed to decompose (very) sparse user-item matrices.

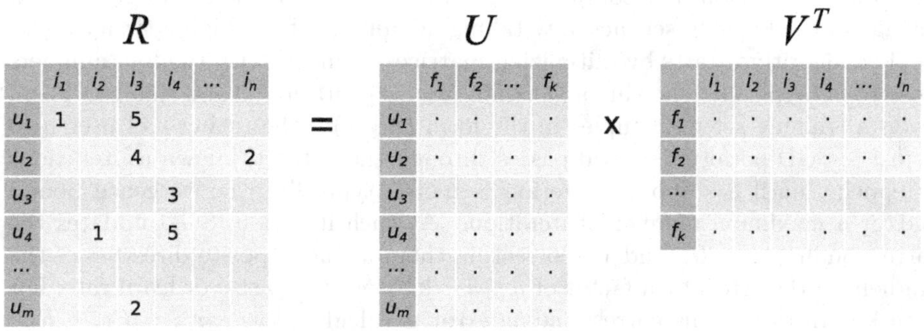

Fig. 1. Matrix factorization: $R = UV^T$.

Figure 1 illustrates the factorization problem. Supposing we have a user-item matrix $R_{m \times n}$ with m users and n items, the algorithm decomposes R in two full factor matrices $U_{m \times k}$ and $V_{n \times k}$ that, similarly to SVD, cover a common k-dimensional latent feature space, such that R is approximated by $\hat{R} = UV^T$. Matrix U spans the user space, while V spans the item space. The k latent features describe users and items in a common space. Given this formulation, the predicted rating by user u to item i is given by a simple dot product:

$$\hat{R}_{ui} = U_u \cdot V_i \tag{5}$$

The number of latent features k is a user given parameter that controls the model size. Naturally, as we increase k, the model relies on a larger number of features, which is beneficial, but only to a certain point. The improvement typically becomes almost unnoticeable when increasing k over a few hundred features.

The model consists of U and V, so the training task consists of estimating the values in U and V that minimize the prediction error on the known ratings. Training is performed by minimizing the L_2-regularized squared error for known values in R:

$$\min_{U.,V.} \sum_{(u,i)\in D} (R_{ui} - U_u \cdot V_i)^2 + \lambda_u ||U_u||^2 + \lambda_i ||V_i||^2 \tag{6}$$

In the above equation, $U.$ and $V.$ are the feature vectors to adjust – corresponding to lines of matrices U and V –, D is the set of user-item pairs for which ratings are known and λ is a parameter that controls the amount of regularization. The regularization terms $\lambda ||U_u||^2$ and $\lambda_i ||V_i||^2$ are used to avoid overfitting. These terms penalize parameters with high magnitudes, that typically lead to overly complex models with low generalization power. For the sake of simplicity, it is common to use $\lambda = \lambda_u = \lambda_i$, which results in a single regularization term $\lambda(||U_u||^2 + ||V_i||^2)$. Stochastic Gradient Descent (SGD) is an efficient method to solve the optimization problem above. It has been informally proposed in [9] and many extensions have been proposed ever since [11,22,26,29]. One obvious advantage of SGD is that complexity grows linearly with the number of known ratings in the training set, actually taking advantage of the high sparsity of R.

The algorithm starts by initializing matrices U and V with random numbers close to 0 – typically following a gaussian $\mathcal{N}(\mu, \sigma)$ with $\mu = 0$ and small σ. Then, given a training set with tuples in the form (u, i, r) – the rating r of user u to item i –, SGD performs several passes through the dataset, known as iterations or epochs, until a stopping criterion is met – typically a convergence bound and/or a maximum number of iterations. At each iteration, SGD updates the corresponding rows U_u and V_i, correcting them in the opposite direction of the gradient of the error, by a factor of $\eta \leq 1$ – known as step size or learn rate. For each known rating, the corresponding error is calculated as $err_{ui} = R_{ui} - \hat{R}_{ui}$, and the following update operations are performed:

$$\begin{aligned} U_u &\leftarrow U_u + \eta(err_{ui}V_i - \lambda U_u) \\ V_i &\leftarrow V_i + \eta(err_{ui}U_u - \lambda V_i) \end{aligned} \tag{7}$$

3.3   Matrix Factorization for Binary Data

The above method is designed to work with ratings. The input of the algorithm is a set of triples in the form (u, i, r), each corresponding to a rating r given by a user u to an item i. It is possible to use the same method with binary data by simply assuming that $r = 1$ for all cases [32]. This results in the following optimization problem:

$$\min_{U.,V.} \sum_{(u,i)\in D} (1 - U_u \cdot V_i)^2 + \lambda_u ||U_u||^2 + \lambda_i ||V_i||^2 \tag{8}$$

In the end, the predicted "rating" $\hat{R}_{ui} = U_u \cdot V_i$ will be a value indicating a user's preference level for an item. This value can be used in a sorting function f to order a list of items:

$$f_{ui} = |1 - \hat{R}_{ui}| \tag{9}$$

In (9), f_{ui} measures the proximity of a predicted rating to 1. U_u and V_i are always adjusted to minimize the error with respect to 1, so it is natural to assume that the most relevant items for a user u are the ones that minimize f. Note that since we are not imposing non-negativity on the model, we need to use the absolute value of the difference in (9).

4   Incrementality

In real world systems, user feedback is generated continuously, at unpredictable rates, and is potentially unbounded – i.e. it is not assumed to stop coming. Moreover, the rate at which user activity data is generated can be very fast. Building predictive models on these continuous flows of data is a problem actively studied in the field of data stream mining [5].

One efficient way to deal with data streams is to maintain incremental models and perform on-line updates as new data points become available. This simultaneously addresses the problem of learning non-stationary concepts and computational complexity issues. However, this requires algorithms able to process data at least as fast as it is generated. Incremental algorithms for recommendation are not frequently addressed in the recommender systems literature, when compared to batch-learning algorithms for recommendation, which are much more thoroughly studied.

User feedback data in recommender systems shares all the characteristics of a data stream. It is continuously generated online, at unpredictable rates and the length of the data is potentially unbounded. Having this in consideration, it becomes clear that the batch approach to recommender systems has fundamental limitations.

Stream mining algorithms should be able to timely process streams, at the risk of not being able to keep up with the arrival rate of data elements. To apply this principle to recommender systems, we simply have to look at the recommendation problem as a data stream problem. This approach has several implications in the algorithms' design and evaluation – see Sect. 7. Regarding the algorithms' design and implementation, one practical way to deal with data streams is to use algorithms that are able to update models incrementally.

4.1   Incremental Neighborhood Methods

Classic neighborhood-based CF algorithms – user- and item-based – have been adapted to work incrementally. The main idea in both cases is to maintain the factors of the similarity function in memory, and update them with simple increments each time a new user-item interaction occurs.

In [21], Papagelis et al. propose an algorithm that incrementally updates the values in the user-user similarity matrix. When a user u rates an item, the similarity values between u and other users are obtained with increments to previous values. Using the Pearson Correlation, the factors of the similarity calculation between user u and another user v are split in the following way:

$$A = \frac{B}{\sqrt{C}\sqrt{D}} \qquad (10)$$

Given the set I of items co-rated by both u and v, factors A, B and C correspond to the following terms:

$A = sim(u, v)$,
$B = \sum_{i \in I}(r_{u,i} - \bar{r}_u)(r_{v,i} - \bar{r}_v)$,
$C = \sum_{i \in I}(r_{u,i} - \bar{r}_u)^2$,
$D = \sum_{i \in I}(r_{v,i} - \bar{r}_v)^2$

It is intended to obtain the new similarity A' from B, C and D, and the new incoming rating:

$$A' = \frac{B'}{\sqrt{C'}\sqrt{D'}} \Leftrightarrow A' = \frac{B+e}{\sqrt{C+f}\sqrt{D+g}} \tag{11}$$

where e, f and g are increments that are easy to calculate after a rating arrives. The actual definitions of e, f and g depend on whether we are adding a new rating or updating an existing one, and we omit them here for the sake of clarity and space. Note that to incrementally update the similarities, the values of B, C and D for all pairs of users must be always available, which requires additional memory. Nevertheless, this simple technique allows fast online updates of the similarity values between the active user and all others.

Miranda and Jorge [18] study incremental algorithms using binary data. The incremental calculation of item-item cosine similarities can be based on user occurrence and co-occurrence counts. A user-user co-occurrence matrix F containing the number of items common to each pair of users can be kept. The diagonal of F contains the number of independent occurrences of each user – i.e. the number of items the user occurs with. Every time a new user-item pair (u, i) is observed in the dataset, the corresponding counts are incrementally updated. Using these counts, the similarities of user u with any other user v can be easily (re)calculated and stored.

$$S_{uv} = sim(u, v) = \frac{F_{uv}}{\sqrt{F_{uu}} \times \sqrt{F_{vv}}} \tag{12}$$

Alternatively, we can use this incremental approach in item-based algorithms.

4.2   Incremental Matrix Factorization

Early work on incremental matrix factorization for recommender systems is presented in [28], where Sarwar et al. propose a method to perform incremental updates of the Singular Value Decomposition (SVD) of the ratings matrix. This is a direct application of the Fold-in method [3], that essentially enables the calculation of new latent vectors (corresponding to new users or new items) based on the current decomposition and by appending them to the corresponding matrices. One shortcoming of this method is that it is applicable only to pure SVD, and it requires an initial batch-trained model.

Takács et al. address the problem of incremental model updates in [29] in a matrix factorization algorithm for ratings data. The idea is to retrain user features every time new ratings are available, but *only* for the active user(s), leaving item features unmodified, avoiding the whole process of batch-retraining the model. This method is a step forward in incremental learning, however it has the following limitations:

- The algorithm requires initial batch training;
- The whole ratings history R is required to update user features;
- Item features are not updated, and new items are not taken into account.

Both the above contributions still require batch learning at some point, that consists of performing several iterations over a learning dataset. While this may be an acceptable overhead in a static environment, it is not straightforwardly applicable with streaming data. As the number of observations increases and is potentially unbounded, repeatedly revisiting all available data eventually becomes too expensive to be performed online.

Fortunately, SGD is *not* a batch algorithm [12]. The only reason why several passes are made over a (repeatedly shuffled) set of data is because there is a *finite* number of examples. Iterating over the examples in different order several times is basically a trick to improve the learning process in the absence of fresh examples. If we assume – as we have to, in a data stream scenario – that there is a continuous flow of examples, this trick is no longer necessary. By this reasoning, SGD can – and should – be used online if enough data is available [2].

In [32], we propose ISGD, an algorithm designed to work as an online process, that updates factor matrices U and V based solely on the current observation. This algorithm, despite its formal similarity with the batch approach, has two practical differences. First, the learning process requires a single pass over the available data – i.e. it is essentially a memoryless algorithm, since there is no fundamental need to revisit past observations. Second, no data shuffling – or any other data pre-processing – is performed. Given that we are dealing with binary feedback we approach the boolean matrix R by assuming the numerical value 1 for *true* values. Accordingly, we measure the error as $err_{ui} = 1 - \hat{R}_{ui}$, and update the rows in A and B^T using the update operations in (7). Since we are mainly interested in top-N recommendation, we need to retrieve an ordered list of items for each user. We do this by sorting candidate items i for each user u using the function $f_{ui} = |1 - \hat{R}_{ui}|$, where $\hat{R}_{ui}$ is the non-boolean predicted score. In plain text, we order candidate items by descending proximity to value 1.

One problem of ISGD is that the absence of negative examples leads to a model that converges globally to the positive class. Take the extreme trivial solution of making $\hat{R}_{ui} = 1$ for all u and i. In this case, $err_{ui} = 0$ in all cases, and no learning would be performed. In a more practical scenario, we would initialize U and V with values close to 0, which would enforce learning. Nevertheless, predictions will accumulate closer and closer around the target positive value. Eventually, the algorithm loses discriminative power, causing accuracy degradation.

We propose a solution for this problem in [33], using recency-based negative feedback imputation. The strategy is to introduce a mechanism that automatically selects *likely* negative examples. The intuition is that the items that have occurred the longest ago in the data stream are better candidates to be taken as negative examples for any user. These are items that no users have interacted with in the longest possible period of activity in the system. We maintain a global priority queue of items occurring in the stream, independently of the user. For every new positive user-item pair (u, i) in the data stream, we introduce a set $\{(u, j_1), \ldots, (u, j_l)\}$ of negative feedback consisting of the active – currently observed – user u and the l items j that are in the tail of the global item queue.

4.2.1   Incremental Learning-To-Rank

Learning to rank encompasses a set of methods that use machine learning to model the precedence of some entities over others, assuming that there is at natural ordering between them. The top-N recommendation task consist of retrieving the best ranked items for a particular user, so it is natural to approach the task as a learn-to-rank problem.

This is the approached followed by Rendle et al. in [24] with their Bayesian Personalized Ranking (BPR) framework. One shortcoming of this algorithm is that it is approached as a batch method. However, although not documented in the literature, the implementation available in the MyMediaLite[1] software library provides an incremental implementation of the algorithm. Essentially, the incremental version of BPR follows the same principle of ISGD, which is to process data points sequentially in one pass, enabling the processing of streaming data.

Another incremental algorithm for ranking that uses a selective sampling strategy is proposed by Diaz-Aviles et al. in [4]. The algorithm maintains a reservoir with a fixed number of observations taken randomly from a stream of positive-only user-item pairs. Every n^{th} pair in the stream is sampled to the reservoir with probability $|R|/n$, with $|R|$ being the number of examples in the reservoir. Model updates are performed by iterating through this reservoir rather than the entire dataset. The strategy is to try to always rely on the most informative examples to update the model.

5   Forgetting

One of the problems of learning from data streams is that the concepts being captured may not be stationary. In recommender systems, users preferences usually change over time. This means that an algorithm that correctly models user preferences in a certain time frame may not accurately represent the same users' preferences some time later. Incremental algorithms benefit from being constantly updated with fresh data, therefore capturing these changes immediately, however the model still retains the concepts learned from past data. One complementary way to deal with this is to forget this outdated information, i.e. data that no longer represent the concept(s) being learned by the algorithm.

---

[1] http://www.mymedialite.net/.

5.1   Forgetting for Neighborhood-Based Incremental CF

In [31], we have used fading factors to *gradually* forget user feedback data using neighborhood-based algorithms. We do this by successively multiplying by a positive scalar factor $\alpha < 1$ all cosine similarities between all pairs of users – or items, in an item-based algorithm – at each incremental step, before updating the similarities with the new observations. If we consider a symmetric similarity matrix S containing all similarity values between pairs of users – or pairs of items –, this is achieved using the update $S \leftarrow \alpha S$. The lower the value of α, the faster the forgetting occurs. In practice, two users – or two items – become less similar as they co-occur farther apart in time.

Our results show that this type of forgetting is beneficial for the algorithms' accuracy, especially in the presence of sudden changes.

5.2   Forgetting with Factorization-Based Incremental CF

We have studied forgetting strategies for incremental matrix factorization in [16]. To achieve forgetting we use a total of eleven forgetting strategies of two types: *rating-based* and *latent-factor-based*. The first performs forgetting of certain past ratings for each user, while the latter performs forgetting by readjusting the latent factors in the user factor matrix, diminishing the impact of past ratings.

Ratings-based forgetting generally consists of forgetting sets of ratings. Formally, it is a function that operates on the set of ratings R_u of a user u and generates a new set $R'_u \subseteq R_u$:

$$f : R_u \rightarrow R'_u$$

We proposed the following six rating-based forgetting methods:

– *Sensitivity-based forgetting*, based on sensitivity analysis. The idea is to forget the ratings that cause changes with higher-than-normal magnitude in the user model. The rationale is that these ratings are typically not representative of the user preferences and should therefore be forgotten. Practical examples of such ratings are the ones on items that are bought as gifts, or when some person uses someone else's account.
– *Global sensitivity-based forgetting*. Like the previous technique, it also forgets ratings that have an impact that falls out of the regular one. The difference is that the sensitivity threshold is measured globally instead of being personalized.
– *Last N retention*. Here the strategy is to retain the latest N ratings for each user. This acts as a sliding window over the ratings of each user with at most N ratings.
– *Recent N retention*. Similar to Last N retention, except that N corresponds to time, instead of a fixed number of ratings, retaining only the ratings that fall into the previous N time units.

- *Recall-based change detection.* This strategy detects sudden drops in the incremental measurement of Recall – i.e. downward variations above a certain threshold, which is maintained incrementally as well – and forgets all ratings occurring before the detected change. This is particularly helpful in environments where changes are abrupt.
- *Sensitivity-based change detection*[2]. This is similar to Recall-based change detection, except that the criterion for detecting a change is the impact of new ratings. If a certain rating changes the user profile dramatically, we assume that the change is real – the user has actually changed preferences – and forget all past ratings.

Latent-factor-based forgetting operates directly on the factorization model, adjusting user or item latent factors in a way that it imposes some type of forgetting. These adjustments to latent factors are linear transformations in the form:

$$A_u^{t+1} = \gamma \cdot A_u^t + \beta$$
$$B_i^{t+1} = \gamma \cdot B_i^t + \beta$$

In the above equations, γ and β are dependent on one of the five strategies below:

- *Forget unpopular items.* This technique consists of penalizing unpopular items by multiplying their latent vectors with a factor proportional to their frequency in the stream.
- *User factor fading.* Here, user latent factors are multiplied by a positive factor $\gamma < 1$. This causes the algorithm to gradually forget user profiles, benefiting recent user activity and penalizing past activity.
- *SD-based user factor fading.* This technique also multiplies user factors by a scalar value, except that this value is not a constant, but rather depends on the volatility of user factors. Users whose factors are more unstable have a higher forgetting rate than those whose profiles are more stable.
- *Recall-based user factor fading.* Similarly to the previous strategy, users have differentiated forgetting factors. This technique amplifies the forgetting factor for users that have low Recall.
- *Forget popular items.* This is the opposite of "Forget unpopular items". Frequent items are penalized as opposed to the non-frequent ones.

Using the BRISMF algorithm [29] as baseline, we implement and evaluate the above strategies on eight datasets, four of which contain positive-only data, while the other four contain numerical ratings.

Our findings in [16] show that forgetting significantly improves the performance of recommendations in both types of data – positive-only and ratings. Latent-factor-based forgetting techniques, and particularly "SD-based user factor fading", have shown to be the most successful ones both on the improvement of recommendations and in terms of computational complexity.

---

[2] The term *sensitivity* is used here with its broader meaning, not as a synonym of recall.

6   Multidimensional Recommendation

The data which are most often available for recommender systems are web access data that represent accesses from users to pages. Therefore, the most common recommender systems focus on these two dimensions. However, other dimensions, such as time and type of content (e.g., type of music that a page concerns in a music portal) of the accesses, can be used as additional information, capturing the context or background information in which recommendations are made in order to improve their performance. For instance, the songs recommended by a music web site (e.g., Last.fm) to a user who likes rock music should be different from the songs recommended to a user who likes pop music.

We can classify the multidimensional recommender systems into three categories [1]: pre-filtering, modeling and post-filtering. In pre-filtering, the additional dimensions are used to filter out irrelevant items before building the recommendation model. Modeling consists in using the additional dimensions inside the recommendation algorithms. In post-filtering, the additional dimensions are used after building the recommendation model to reorder or filter out recommendations.

One approach that combines pre-filtering and modeling, called **DaVI** (*Dimensions as Virtual Items*), enables the use of multidimensional data in traditional two-dimensional recommender systems. The idea is to treat additional dimensions as virtual items, using them together with the regular items in a recommender system [7]. Virtual items are used to build the recommendation model but they can not be recommended. On the other hand, regular items are used to build the model and they can also be recommended. This simple approach provides good empirical results and allows the use of existing recommenders.

The **DaVI** approach consists in converting each multidimensional session into an extended two-dimensional session. The values of the additional dimensions, such as "day = monday" are used as virtual items together with the regular items. The **DaVI** approach can also be applied to a subset of dimensions or even to a single dimension.

Once we convert a set of multidimensional sessions S' into a set of extended two-dimensional sessions S'', building/learning a multidimensional recommendation model consists in applying a two-dimensional recommender algorithm on S''. We illustrate the learning process using the **DaVI** approach in Fig. 2, where the values of the additional dimension *Hour* are used as virtual items.

7   Evaluation

There are two main categories of evaluation procedures: offline and online. In the latter case, the recommender system is run on real users, and typically an A/B test is conducted [6]. In the former case the system is evaluated on archived data, where part is saved for training a model and the rest for testing the result. Offline evaluation is arguably not enough for assessing the power of recommendations [17]. Moreover, there is no guarantee that an algorithm with

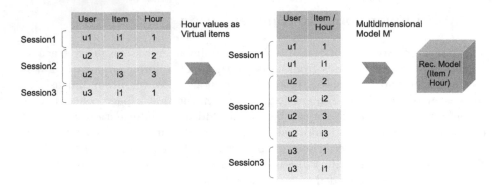

Fig. 2. Illustration of the learning process using the **DaVI** approach.

good offline results will have good online performance, from the users' perspective [23]. However, offline evaluation is important to assess the predictive ability and runtime performance of algorithms. It also has the advantages of being cheaper than online evaluation and of enabling repeatability.

7.1   Offline Evaluation Methodologies

Offline evaluation refers to evaluation methodologies using previously collected datasets, and conducted in a controlled laboratory environment, with no interaction with users. Offline protocols allow researchers to evaluate and compare algorithms by simulating user behavior. In the recommender systems literature, this typically begins by splitting the dataset in two subsets, the training set and testing set, picking random data elements from the initial dataset. The training set is used to train the recommender model. Evaluation is done by comparing the predictions of the model with the actual data in the test subset. Different protocols can be used. Generally, these protocols group the test set by user – or user session – and "hide" user-item interactions randomly chosen from each group. These hidden interactions form the hidden set. Rating prediction algorithms are usually evaluated by comparing predicted ratings with the hidden ratings. Item recommendation algorithms are evaluated performing user-by-user comparison of the recommended items with the hidden set.

Offline evaluation protocols are usually easy to implement, and enable the reproducibility of experiments, which is an important factor in peer-reviewed research. However they are typically designed to work with static models and finite datasets, and are not trivially applicable to streaming data and incremental models.

7.1.1   Prequential Evaluation

To solve the issue of how to evaluate algorithms that continuously update models, we have proposed a prequential methodology [10] to evaluate recommender systems. Evaluation is made treating incoming user feedback as a data stream.

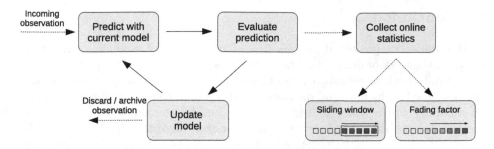

Fig. 3. Prequential evaluation

Evaluation is continuously performed in a test-then-learn scheme (Fig. 3): whenever a new rating arrives, the corresponding prediction is scored according to the actual rating. This new rating is then used to update the model.

Take the example of a top-N recommendation task with binary data. In a binary data stream, observations do not contain actual ratings. Instead, each observation consists of a simple user-item pair (u, i) that indicates a positive interaction between user u and item i. The following steps are performed in the prequential evaluation process:

1. If u is a known user, use the current model to recommend a list of items to u, otherwise go to step 3;
2. Score the recommendation list given the observed item i;
3. Update the model with (u, i) (optionally);
4. Proceed to – or wait for – the next observation.

One important note about this process is that it is entirely applicable to algorithms that learn either incrementally or in batch mode. This is the reason why step 3 is annotated as optional. For example, instead of performing this step, the system can store the data to perform batch retraining periodically.

This protocol provides several benefits over traditional batch evaluation:

– It allows continuous monitoring of the system's performance over time;
– Several metrics can be captured simultaneously;
– If available, other kinds of user feedback can be included in the loop;
– Real-time statistics can be integrated in the algorithms' logic – e.g. automatic parameter adjustment, drift/shift detection, triggering batch retraining;
– In ensembles, relative weights of individual algorithms can be adjusted;
– The protocol is applicable to both positive-only and ratings data;
– Offline experiments are trivially reproducible if the same data is available.

In an offline experimental setting, an overall average of individual scores can be computed at the end – because offline datasets are finite – and on different time horizons. For a recommender running in a production system, this process allows us to follow the evolution of the recommender by keeping online statistics of the metrics (e.g. a moving average of accuracy, or an error rate). Thereby it is possible to depict how the algorithm's performance evolves over time.

8 Conclusion

In this paper we presented an overview of recommender systems mostly centered on the work of our team. The main focus is on incremental approaches for binary data both to neighbourhood based and matrix factorization algorithms. We presented an incremental distance based collaborative algorithm where only an auxiliary co-frequency matrix is cached, besides the similarity matrix. We have also described a Stochastic Gradient Descent algorithm for binary matrix factorization that exploits negative feedback. A number of forgetting techniques that cope with the natural dynamics of continuously arriving data were mentioned. Another line of research is on multidimensional approaches, where we proposed the use of virtual items. Finally, we have described an offline prequential evaluation methodology adequate to incremental approaches.

Acknowledgements. (FourEyes) Project "NORTE-01-0145-FEDER-000020" is financed by the North Portugal Regional Operational Programme (NORTE 2020), under the PORTUGAL 2020 Partnership Agreement, and through the European Regional Development Fund (ERDF).

References

1. Adomavicius, G., Tuzhilin, A.: Context-aware recommender systems. In: Pu, P., Bridge, D.G., Mobasher, B., Ricci, F. (eds.) Proceedings of the 2008 ACM Conference on Recommender Systems, Rec-Sys 2008, Lausanne, 23–25 October 2008, pp. 335–336 (2008)
2. Bottou, L.: Stochastic learning. In: Bousquet, O., Luxburg, U., Rätsch, G. (eds.) ML -2003. LNCS (LNAI), vol. 3176, pp. 146–168. Springer, Heidelberg (2004). doi:10.1007/978-3-540-28650-9_7
3. Deerwester, S.C., Dumais, S.T., Landauer, T.K., Furnas, G.W., Harshman, R.A.: Indexing by latent semantic analysis. JASIS **41**(6), 391–407 (1990)
4. Diaz-Aviles, E., Drumond, L., Schmidt-Thieme, L., Nejdl, W.: Real-time top-N recommendation in social streams. In: Cunningham, P., Hurley, N.J., Guy, I., Anand, S.S. (eds.) Sixth ACM Conference on Recommender Systems, RecSys 2012, Dublin, 9–13 September, pp. 59–66. ACM (2012)
5. Domingos, P.M., Hulten, G.: Mining high-speed data streams. In: Ramakrishnan, R., Stolfo, S.J., Bayardo, R.J., Parsa, I. (eds.) Proceedings of the Sixth ACM SIGKDD International Conference on Knowledge Discovery and Data Mining, Boston, 20–23 August, pp. 71–80. ACM (2000)
6. Domingues, M.A., Gouyon, F., Jorge, A.M., Leal, J.P., Vinagre, J., Lemos, L., Sordo, M.: Combining usage, content in an online recommendation system for music in the long tail. IJMIR **2**(1), 3–13 (2013)
7. Domingues, M.A., Jorge, A.M., Soares, C.: Dimensions as virtual items: improving the predictive ability of top-N recommender systems. Inf. Process. Manag. **49**(3), 698–720 (2013)
8. Duan, L., Street, W.N., Xu, E.: Healthcare information systems: data mining methods in the creation of a clinical recommender system. Enterp. Inf. Syst. **5**(2), 169–181 (2011)

9. Funk, S.: http://sifter.org/~simon/journal/20061211.html (2006). Accessed Jan 2013
10. Gama, J., Sebastião, R., Rodrigues, P.P.: On evaluating stream learning algorithms. Mach. Learn. **90**(3), 317–346 (2013)
11. Koren, Y.: Factorization meets the neighborhood: a multifaceted collaborative filtering model. In: Li, Y., Liu, B., Sarawagi, S. (eds.) Proceedings of the 14th ACM SIGKDD International Conference on Knowledge Discovery and Data Mining, Las Vegas, 24–27 August, pp. 426–434. ACM (2008)
12. LeCun, Y.A., Bottou, L., Orr, G.B., Müller, K.-R.: Efficient backprop. In: Montavon, G., Orr, G.B., Müller, K.-R. (eds.) Neural Networks: Tricks of the Trade. LNCS, vol. 7700, pp. 9–48. Springer, Heidelberg (2012). doi:10.1007/978-3-642-35289-8_3
13. Lee, D., Hosanagar, K.: When do recommender systems work the best? The moderating effects of product attributes and consumer reviews on recommender performance. In: Bourdeau, J., Hendler, J., Nkambou, R., Horrocks, I., Zhao, B.Y. (eds.) Proceedings of the 25th International Conference on World Wide Web, WWW, Montreal, 11–15 April, pp. 85–97. ACM (2016)
14. Linden, G., Smith, B., York, J.: Amazon.com recommendations: item-to-item collaborative filtering. IEEE Internet Comput. **7**(1), 76–80 (2003)
15. Manouselis, N., Drachsler, H., Verbert, K., Duval, E.: Recommender Systems for Learning. Springer Briefs in Electrical and Computer Engineering. Springer, New York (2013)
16. Matuszyk, P., Vinagre, J., Spiliopoulou, M., Jorge, A.M., Gama, J.: Forgetting methods for incremental matrix factorization in recommender systems. In: Wainwright et al. [35], pp. 947–953
17. McNee, S.M., Riedl, J., Konstan, J.A.: Being accurate is not enough: how accuracy metrics have hurt recommender systems. In: Olson, G.M., Jeffries, R. (eds.) Extended Abstracts Proceedings of the Conference on Human Factors in Computing Systems, CHI, Montréal, 22–27 April, pp. 1097–1101. ACM (2006)
18. Miranda, C., Jorge, A.M.: Incremental collaborative filtering for binary ratings. In: 2008 IEEE/WIC/ACM International Conference on Web Intelligence, WI 2008, 9–12 December 2008, Sydney, Main Conference Proceedings, pp. 389–392. IEEE Computer Society (2008)
19. Moreira-Matias, L., Fernandes, R., Gama, J., Ferreira, M., Mendes-Moreira, J., Damas, L.: On recommending urban hotspots to find our next passenger. In: Gama, J., May, M., Marques, N.C., Cortez, P., Ferreira, C.A. (eds.) Proceedings of the 3rd Workshop on Ubiquitous Data Mining Co-located with the 23rd International Joint Conference on Artificial Intelligence (IJCAI), Beijing, 3 August. CEUR Workshop Proceedings, vol. 1088, p. 17. CEUR-WS.org (2013)
20. Nguyen, T.T., Hui, P.-M., Harper, F.M., Terveen, L., Konstan, J.A.: Exploring the filter bubble. In: Proceedings of the 23rd International Conference on World Wide Web - WWW 2014, pp. 677–686 (2014)
21. Papagelis, M., Rousidis, I., Plexousakis, D., Theoharopoulos, E.: Incremental collaborative filtering for highly-scalable recommendation algorithms. In: Hacid, M.-S., Murray, N.V., Raś, Z.W., Tsumoto, S. (eds.) ISMIS 2005. LNCS (LNAI), vol. 3488, pp. 553–561. Springer, Heidelberg (2005). doi:10.1007/11425274_57
22. Paterek, A.: Improving regularized singular value decomposition for collaborative filtering. In: Proceedings of KDD Cup and Workshop 2007, pp. 5–8 (2007)
23. Pearl, P., Chen, L., Rong, H.: Evaluating recommender systems from the user's perspective: survey of the state of the art. User Model. User Adapt. Interact. **22**(4), 317–355 (2012)

24. Rendle, S., Freudenthaler, C., Gantner, Z., Schmidt-Thieme, L.: BPR: Bayesian personalized ranking from implicit feedback. In: Bilmes, J.A., Ng, A.Y. (eds.) Proceedings of the Twenty-Fifth Conference on Uncertainty in Artificial Intelligence, UAI 2009, Montreal, 18–21 June, pp. 452–461. AUAI Press (2009)
25. Ricci, F., Rokach, L., Shapira, B., Kantor, P.B. (eds.): Recommender Systems Handbook. Springer, New York (2011)
26. Salakhutdinov, R., Mnih, A.: Probabilistic matrix factorization. In: Platt, J.C., Koller, D., Singer, Y., Roweis, S.T. (eds.) Advances in Neural Information Processing Systems 20, Proceedings of the Twenty-First Annual Conference on Neural Information Processing Systems, Vancouver, 3–6 December, pp. 1257–1264. Curran Associates Inc. (2007)
27. Sarwar, B.M., Karypis, G., Konstan, J.A., Riedl, J.: Item-based collaborative filtering recommendation algorithms. In: Shen, V.Y., Saito, N., Lyu, M.R., Zurko, M.E. (eds.) Proceedings of the Tenth International World Wide Web Conference, WWW 2010, Hong Kong, 1–5 May, pp. 285–295. ACM (2001)
28. Sarwar, B.M., Karypis, G., Konstan, J.A., Riedl, J.T.: Application of dimensionality reduction in recommender system - a case study. In: WEBKDD 2000: Web Mining for E-Commerce - Challenges and Opportunities, 20 August, Boston (2000)
29. Takács, G., Pilászy, I., Németh, B., Tikk, D.: Scalable collaborative filtering approaches for large recommender systems. J. Mach. Learn. Res. **10**, 623–656 (2009)
30. Veloso, M., Jorge, A., Azevedo, P.J.: Model-based collaborative filtering for team building support. ICEIS (2), 241–248 (2004)
31. Vinagre, J., Jorge, A.M.: Forgetting mechanisms for scalable collaborative filtering. J. Braz. Comput. Soc. **18**(4), 271–282 (2012)
32. Vinagre, J., Jorge, A.M., Gama, J.: Fast incremental matrix factorization for recommendation with positive-only feedback. In: Dimitrova, V., Kuflik, T., Chin, D., Ricci, F., Dolog, P., Houben, G.-J. (eds.) UMAP 2014. LNCS, vol. 8538, pp. 459–470. Springer, Heidelberg (2014). doi:10.1007/978-3-319-08786-3_41
33. Vinagre, J., Jorge, A.M., Gama, J.: Collaborative filtering with recency-based negative feedback. In: Wainwright et al. [34], pp. 963–965
34. Wainwright, R.L., Corchado, J.M., Bechini, A., Hong, J. (eds.): Proceedings of the 30th Annual ACM Symposium on Applied Computing, Salamanca, 13–17 April. ACM (2015)

User Control in Recommender Systems: Overview and Interaction Challenges

Dietmar Jannach, Sidra Naveed, and Michael Jugovac[✉]

Department of Computer Science, TU Dortmund, Dortmund, Germany
{dietmar.jannach,sidra.naveed,michael.jugovac}@tu-dortmund.de

Abstract. Recommender systems have shown to be valuable tools that help users find items of interest in situations of information overload. These systems usually predict the relevance of each item for the individual user based on their past preferences and their observed behavior. If the system's assumption about the users' preferences are however incorrect or outdated, mechanisms should be provided that put the user into control of the recommendations, e.g., by letting them specify their preferences explicitly or by allowing them to give feedback on the recommendations. In this paper we review and classify the different approaches from the research literature of putting the users into active control of what is recommended. We highlight the challenges related to the design of the corresponding user interaction mechanisms and finally present the results of a survey-based study in which we gathered user feedback on the implemented user control features on Amazon.

1 Introduction

Recommender systems have become an integral part of many commercial websites like Amazon, Netflix, and YouTube. In scenarios where millions of choices are available these systems serve as an aid for users in their search and decision making processes by automatically assessing the users' preferences and by making personalized recommendations.

On many websites, including the above-mentioned ones, the underlying user preference model is established by the system by observing and interpreting the users' behavior over time ("implicit feedback") or by considering the user's explicit ratings for individual items. The estimated preference model is then used to make predictions about the relevance of each recommendable item for the user. Over the last two decades a variety of algorithms was proposed in the literature to optimize these relevance assessments using datasets that represent a snapshot of the user's preferences.

In reality, however, the user interests can change over time, which means that some preference information can become outdated, leading to inaccurate recommendations [1,2]. Furthermore, the relevance of an item can depend on the user's current situation. The user might, for example, be looking for a gift which does not fall into his or her typical preference profile. Or, the user might

© Springer International Publishing AG 2017
D. Bridge and H. Stuckenschmidt (Eds.): EC-Web 2016, LNBIP 278, pp. 21–33, 2017.
DOI: 10.1007/978-3-319-53676-7_2

have just bought a certain item so that further recommending the same or similar objects becomes pointless.

In many real-world recommender systems users have limited or no means to inform the system that its assumptions are incorrect or to specify that preference information has become outdated.[1] Past research has however shown that at least in some application domains users appreciate being more actively involved in the process and in control of their recommendations [3,4]. In the end, providing additional forms of user interactions can not only lead to higher user satisfaction but also increase the users' trust in the system [5,6].

In the research literature, a number of proposals have been made on how to implement mechanisms for increased user control. Simple approaches are, for example, based on static preference forms. Others use conversational dialogs or critiquing mechanisms to let the users specify their constraints and preferences. Some proposals even allow the user to choose between different recommendation strategies. Generally, the proposed mechanisms provide different levels of user control but they unfortunately all come with their own challenges regarding the user interaction.

In this paper we first provide an overview of user control mechanisms from the literature, categorize them according to their context in the recommendation process, and discuss the individual user interaction challenges. As a case study of a real system, we then report the findings of a survey-based study in which we investigated how users perceive the comparably powerful explanation, feedback, and control mechanisms that are implemented on Amazon's website. Our observations indicate that although the implemented features are known to many study participants, most users are hesitant to use the provided functionality for different reasons.

2   Control Mechanisms and User Interaction Challenges

2.1   Conceptual Framework

Overview. Figure 1 shows an overview of the approaches and situations in the recommendation process where users can be put into control according to the literature. We categorize the different techniques in two classes:

- Techniques where users are allowed to explicitly specify their preferences. These will be discussed in Sect. 2.2.
- Techniques that put the user into control in the context of recommendation results. We review these approaches in Sect. 2.3.

Critiquing-based techniques share characteristics of both categories. We will discuss them also in Sect. 2.2.

---

[1] In some rating-based systems users can update their ratings, which might however be tedious, and changes often have no immediate effect on the presented recommendations.

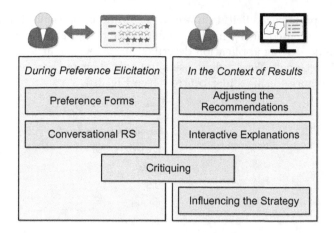

Fig. 1. User control mechanisms for recommender systems

Definition of User Control. In the context of this work we require that user control mechanisms have an *immediate effect* on the recommendations. For example, selecting a certain interest category in a preference form should immediately lead to updated results the next time the recommendations are displayed. Changing or adding explicit item ratings therefore do not count as control mechanisms as usually these changes are not immediately reflected in the results, e.g., because the trained models are only updated periodically. The same holds for like/dislike buttons, which some websites display for each recommendation, in case these have no immediate effect on the next recommendations.

Our second, but softer, requirement is that users should understand or at least have a confident intuition about the effects of their control actions. A "thumbs down" action for the currently played track on the music service Spotify for example results in an immediate update of the next tracks to be played. The logic behind the update is however not transparent, which is why we consider this as a limited form of user control.

Finally, control in recommender systems is sometimes discussed in the literature together with "inspectability", e.g., in [3]. Inspectability, i.e., giving the user insights on what the recommendations are based on, is in our view not a requirement for a control mechanism but can be useful to help users understand the possible effects of their control actions.

2.2 Control During the Preference Elicitation Phase

Preference Forms and Adaptive Dialogs. One basic option of giving control to the users is to let them specify their constraints and preferences explicitly by using *static user profile forms*. Figure 2a shows a simple screen that allows users to choose their genre interest for the Netflix movie recommender. In some applications, such preference forms are used to indirectly infer the interests, e.g., by asking the users for their favorite movies or artists. Such simple forms of user control during preference elicitation are for example implemented in the music recommender presented in [7], in the MenuMentor restaurant menu system [8], and in the energy-saving application described in [9]. A similar approach is also implemented on Google News, where users indicate their preferences about news in different categories using slider controls (see Fig. 2b).

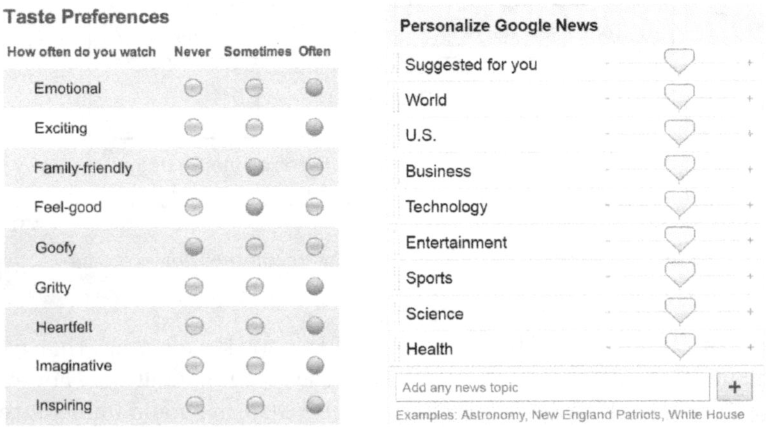

Taste Preferences

How often do you watch	Never	Sometimes	Often
Emotional	◉	◉	◉
Exciting	◉	◉	◉
Family-friendly	◉	◉	◉
Feel-good	◉	◉	◉
Goofy	◉	◉	◉
Gritty	◉	◉	◉
Heartfelt	◉	◉	◉
Imaginative	◉	◉	◉
Inspiring	◉	◉	◉

Personalize Google News

Suggested for you
World
U.S.
Business
Technology
Entertainment
Sports
Science
Health

Add any news topic   +
Examples: Astronomy, New England Patriots, White House

a) *Taste Preference Selection on the Netflix Movie Streaming Site*

b) *Preference Indicators on the Google News Site*

Fig. 2. Static preference forms for personalized services

Generally, static forms are comparably easy to use. However, as soon as the user is allowed to indicate *relative preferences*, these forms can become complicated in terms of their interpretation. For example, in case of the Google News preference indicators it is not clear if having all sliders in the middle position has the same meaning as having all at the maximum level. Another problem with such static forms is that every time the users' interests change, they have to manually adapt their settings such that they properly reflect their new interests.

Because static forms are identical for all users, they might not be optimally suited to capture the preferences of all kinds of users, who can have different levels of expertise in a domain. *Conversational approaches* in some sense try to mimic a human advisor for high-involvement products like digital cameras, e.g., by guiding the user through an interactive dialog based on desired functional features or by providing additional explanations when requested. An early system is the ADAPTIVE PLACE ADVISOR [10], which, according to the classification in [11], adapts its conversation behavior to the users at the information filtering and navigation levels. Similar ideas are implemented in the ADVISOR SUITE system [12], which also adapts the conversation based on the user's previous answers and in addition is capable of explaining the recommendations and can help users in situations in which no item fulfills all their requirements.

Technically, these conversational systems often implement item filtering rules that deterministically map functional features to technical product characteristics. Users of such systems are therefore in immediate control of the recommendation outcomes. Implementing such systems can however require significant knowledge engineering efforts to encode and maintain the recommendation rules. Usually, these systems also do not learn over time or from the behavior of a larger community. From an interaction perspective, users can also feel overwhelmed when they try to change some of their specifications after the initial conversational elicitation phase.

Critiquing. Similar to the discussed form-based techniques, users of *critiquing* approaches explicitly state their preferences on certain item features. Here, however, they do that in the context of a reference item, e.g., a camera, and the provided preferences are *relative* statements like "cheaper" or "higher resolution". The system then uses this feedback to find other items that fulfill the refined requirements. The process is repeated until the user finds a suitable camera or gives up on the search. Critiquing based systems were presented, e.g., in [8,13,14], and a number of works have been proposed to improve the basic interaction scheme, including *compound* or *dynamic* critiques, where users can for example update their preferences in more than one dimension at a time.

Critiquing approaches have the advantage that their general operating principle is easy to understand for the users. Furthermore, each interaction is followed by an immediate update of the recommendation(s). However, basic critiquing schemes can lead to a high number of required iterations until a suitable product is found. Compound critiques, on the other hand, can induce higher cognitive load for the users. Finally, similar to form-based approaches the problem can arise that no more suitable items remain that can be recommended, which means that the system has to implement a recovery strategy for the user.

2.3   Control in the Context of Recommendation Results

Dynamically Adjusting the Recommendations. Once a set of recommendations is computed, a simple form of allowing users to influence what is presented is to provide them with mechanisms to further filter and re-sort the items based on their features. Such a post-filtering functionality was for instance implemented in the MovieCritic [15] and the MetaLens systems [16], where users could for example include or exclude movies of certain genres. In the MetaLens system, users could also indicate the relative importance of individual features.

A more sophisticated and visually complex approach was proposed in [17]. Their system displays three pieces of information in parallel – the items that the user has rated, the corresponding set of similar users, and the resulting recommendations. Users can then not only inspect why certain items were recommended but also interactively adapt their ratings, which is then reflected in updated recommendations.

TasteWeights [18] is a similar approach that also combines a visualization of the recommendation logic with an interaction mechanism. Their system presents a graph that shows the relationships between the user's rated items, their social friends, and the recommendations. The implemented control mechanism allows users to adjust the weights of the items and the friends. Similar to the work in [17], an experimental evaluation indicates that such a form of user involvement can lead to higher user satisfaction.

Another comparable approach was proposed in [19], where a web-based interactive visualization for a content recommender system for microblogs was devised. Their interface also consists of three columns and users can for example change the sort criterion of the items (tweets) or vary the relative importance of different filters.

Overall, all of the presented approaches to put users into control lead to immediate effects on the resulting recommendations. In most cases, the users will at least to some extent understand how their actions (indirectly) impact the outcomes. However, one cannot generally assume that average users will understand the underlying rationale of, e.g., a neighborhood based method. A limitation of some of the works is in fact that they employ comparably simple recommendation methods, and it is unclear how such approaches would work for more complex machine learning models. In addition, users might have rated dozens of items over time and might have a large neighborhood so that manipulating ratings or weights on such a fine-grained level might soon become tedious.

User Control in the Context of Explanations. The literature suggests that providing explanations for recommendations can be beneficial in different ways as they, for example, help users understand why certain items were recommended. This in turn could lead to increased satisfaction and trust in the system [20]. Since mechanisms for user control often require that users understand or at least have an intuition of the reasoning logic of the system, designing these mechanisms in the context of explanations appears natural.

In the context of conversational systems, such interactive explanations were for example developed for the ADVISOR SUITE system described above [21]. The knowledge-based system was able to generate textual explanations based on the logical rules that map user preferences to item characteristics. In case some requirements could not be fulfilled, users were able to overwrite the default priorities of the rules with their personal preference weights. This feedback was then immediately processed to compute an updated recommendation list.

Feedback mechanisms in the context of explanations were also implemented in the mobile recommender system Shopr [22]. In this system the explanations were provided along with the recommendations, e.g., *"Because you were interested in ... in the past"*, and users could then give feedback to the system about whether this recommendation logic was inappropriate in their current situation. A possible type of feedback included not only to indicate that a certain item is not relevant but a whole category of items should not be recommended. A similar feature, although not in the context of explanations, can be found on YouTube, see Fig. 3.

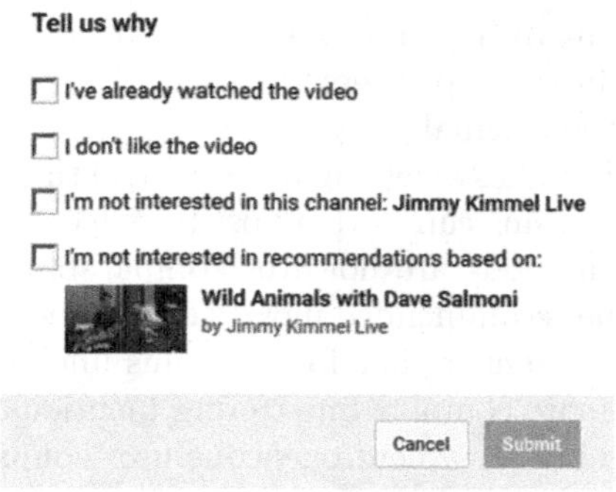

Tell us why

☐ I've already watched the video

☐ I don't like the video

☐ I'm not interested in this channel: **Jimmy Kimmel Live**

☐ I'm not interested in recommendations based on:

Wild Animals with Dave Salmoni
by Jimmy Kimmel Live

Cancel   Submit

Fig. 3. Feedback options for a recommendation on the YouTube platform.

The discussed methods in general allow users to correct possibly wrong assumptions in the context of what is sometimes called "scrutable" interactive explanations [23]. The concept of these scrutable explanations is that with their help users are able to inspect and understand (scrutinize) the system's reasoning and act upon this knowledge to improve the system's assumptions [20].

As with all forms of user control discussed so far, it can be challenging to design such interactive explanations when the underlying reasoning mechanisms are complex. In these cases, generating understandable explanations can represent a problem of its own. In Sect. 3, we will discuss the interactive explanation mechanism that is implemented on Amazon in more detail.

Choosing or Influencing the Recommendation Strategy. A quite different approach of letting users influence the recommendations is to allow them to select or parameterize the algorithms themselves that are used to generate the recommendations. In the study described in [24], for example, users of the MovieLens system were able to choose one of four predefined algorithms by clicking on a widget in the top menu bar. Each selection immediately led to a different set of recommendations. An analysis of the log files revealed that about one quarter of the users actually tried out the recommendation-switching feature.

More fine-grained control was given to users in the approach presented in [25], where users could fine-tune the importance weights of a hybrid recommender. Their graphical interface furthermore visualized – with the help of a Venn diagram – based on which algorithm each item was included in the recommendations. Two user studies were performed to assess the effect of the system on the users and the authors report that their system led to higher user engagement, and it furthermore seems that users worked more efficiently with the tool.

User control in terms of interactively fine-tuning the desired item characteristics was recently proposed and experimentally analyzed in [26]. The participants of a user study could for example change the popularity or recency level of the movies to be recommended. When the results were presented, the users could then use additional knobs to fine-tune the results until they were satisfied. An analysis revealed that, in the end, users were happier with the adapted recommendations than with the original ones.

Overall, the different studies of letting users control the underlying algorithms indicate that such mechanisms can have a positive effect on the user experience. Some of the proposed methods are however comparably simple. Letting users vary the popularity of the recommended movies can be seen as a form of changing the sort order. To some extent it therefore remains unclear how user interfaces should be designed for more complex fine-tuning functionalities as they have to be intuitive and easy to use for a heterogeneous user community.

3   On the Acceptance of Amazon's Scrutable Explanations

Our literature overview in general indicates that finding appropriate user interface mechanisms for putting users into control can be challenging. In the end, a poor UI design can lead to limited acceptance of the control mechanisms by the users, e.g., because they find the system tedious or have problems understanding the effects of their actions.

To obtain a better understanding of mechanisms for user control in the context of recommendations, we conducted a user survey about the acceptance of the comparably rich explanation, feedback, and control functionality that is implemented on the websites of Amazon. Specifically, our goals were to assess to what extent users are aware of the functionality, if they find the provided functionality clear, and if they are actually using it.

User feedback and control on Amazon is provided in the context of explanations as proposed in the literature in [21, 22] or [23]. Figure 4 shows a screenshot of the provided functionality.

The presented screen can be reached by navigating to dedicated pages on which users can inspect and improve their recommendations. For some product categories the functionality can be accessed directly from the recommendation lists, but Amazon seems to vary the UI design in that respect over time.

Fig. 4. Explanation-based control mechanism at Amazon

In Fig. 4 a recommended item is displayed along with an explanation why it is recommended. The explanation in this case refers to another item that the user already owns. The user can then give feedback on the recommended item by rating it or indicating no interest in the item. Furthermore, the users can instruct the system not to use the already owned item for future recommendations. This functionality can for example be helpful when the user's interest has changed over time or when the item is outside of his or her usual interests, e.g., because it was a gift.

3.1 Survey Design

We conducted a two-phase survey-based study in which we asked users of Amazon about their knowledge and usage behavior regarding the feedback mechanisms implemented on the site. All participants were computer science students of our university and at the same time regular customers of Amazon doing several purchases a year.

Phase 1: 75 students participated in the first part. The survey sheet showed screenshots of Amazon's functionality and comprised 15 questions with 5-point Likert-scale answer options. Questions were for example about whether or not the participants know the functionality, if they find the functionality clear, and whether or not they have used or intend to use it in the future.[2]

Phase 2: The second phase, which took place a few weeks later, particularly focused on reasons why the users would or would *not* click on the recommendations and on possible reasons for *not using* the feedback functionality. 26 students of the user population from the first phase returned the survey sheets, which – besides a set of Likert-scale questionnaire items – included two free-text fields where the participants could express reasons *not to use* the provided functionalities.

3.2 Observations

Phase 1. A first surprising observation is that more than 75% of the participants stated that they use recommendations on the site never or only rarely. At the same time, they found it on average "rather clear" or "very clear" how the recommendations are created. The average answer value was 4.04 on the five-point scale, where five means "very clear".

When asked whether they knew that they could influence their recommendations, more than 90% of the subjects answered positively, even though not all of them knew exactly how. However, only about 20% were aware of the special page for improving recommendations and even fewer had ever used the page.

Regarding the feature called *"Don't use for recommendations"*, more than 50% stated that the provided functionality was clear or very clear to them.

---

[2] A translated version of the survey forms can be found at
http://ls13-www.cs.tu-dortmund.de/homepage/publications/ec-web-2016/.

Another 25% said that one could guess its purpose. On the other hand, only about 8% of the users (6 out of 76) had ever actually used the functionality.

We then asked the participants whether they had ever noticed the provided explanation (*"Because you bought ..."*). Around 40% answered with yes. Again, the majority of subjects stated that the functionality is mostly clear to them, but only 4 of 76 had ever used the *"Don't use for recommendation"* feature in that context. Finally, although the participants found the functionality potentially useful (avg. answer was 3.36 on the five-point scale, where five represents the most positive answer), the intention to use the feature in the future was rather limited (2.55).

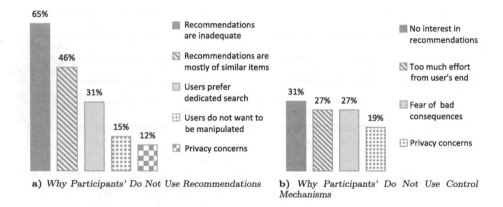

a) *Why Participants' Do Not Use Recommendations*

b) *Why Participants' Do Not Use Control Mechanisms*

Fig. 5. Results of the Amazon survey

Phase 2. In the second phase we were particularly interested in the reasons why the participants are hesitant to use the provided recommendation, feedback and control mechanisms. We manually analyzed the qualitative free-form feedback from the subjects and categorized the responses in different groups.

Figure 5a shows the statistics for the reasons why many of the participants would not use Amazon's recommendations. The two main reasons are related to the quality of the recommendations, which appear not to be very helpful or contain too many similar items.[3] Another common reason for not relying on recommendations is that users prefer to use explicit search. Finally, privacy concerns and fear of being manipulated were other aspects mentioned by the participants in this survey.

In Fig. 5b we summarize the reasons for not using the *"Don't use for recommendations"* feature. One main reason is that the participants do not use the recommendations in the first place. Many however also found that this form of fine-tuning requires too much effort. An equal amount of respondents were afraid of the consequences of their actions and of the inability to undo their settings later on. A smaller amount of participants again mentioned privacy issues.

---

[3] The participants could provide several reasons and the value 65% indicates that nearly two thirds of the users stated that the recommendations were inadequate.

3.3 Discussion

Our results indicate that although the mechanisms provided on Amazon were known to many participants, they are hesitant to actually use them, e.g., due to the additional effort or unclear consequences of their actions.

Furthermore, the responses of the users in general indicate limited satisfaction and trust in the recommendation and feedback system. Providing mechanisms that are understandable for users and have an immediate effect on the recommendations seems to be required but not sufficient, which calls for better mechanisms to put users into control. More user-friendly systems could, for example, provide less tedious forms of interaction or clearly indicate that profile changes can be undone to reduce the users' fear of undesired consequences.

Overall, we see our survey as a further step toward a better understanding of user control mechanisms for recommenders. However, a number of questions remains open for future research. Further studies could, for example, continue related lines of research described in [9, 27–30] and further investigate what level of control users expect from recommender systems, whether more control is always better, or if different users call for different control mechanisms. Also, further research is necessary to identify the effects of control mechanisms on user engagement and the user experience.

Regarding research limitations, note that our survey is based on the responses of computer science students, who might have a representative online shopping behavior for their age group but are maybe untypical in some aspects, e.g., with respect to privacy concerns. The sample size of the initial study reported in this paper also represents a research limitation.

4 Conclusions

We have reviewed the literature on user control in recommender systems and have identified different requirements to make such approaches effective in particular with respect to the design of the user interaction mechanisms. A survey among users of Amazon indicates that the provided functionality is only used to a very limited extent. Besides the poorly regarded quality of the recommender system itself, the major reasons include the partially unclear consequences of feedback and control actions and the increased user effort, which indicates that more research is required in this area.

References

1. Hu, Y., Koren, Y., Volinsky, C.: Collaborative filtering for implicit feedback datasets. In: ICDM 2008, pp. 263–272 (2008)
2. Amatriain, X., Pujol, J.M., Tintarev, N., Oliver, N.: Rate it again: increasing recommendation accuracy by user re-rating. In: RecSys 2009, pp. 173–180 (2009)
3. Knijnenburg, B.P., Bostandjiev, S., O'Donovan, J., Kobsa, A.: Inspectability and control in social recommenders. In: RecSys 2012, pp. 43–50 (2012)

4. Dooms, S., De Pessemier, T., Martens, L.: Improving IMDb movie recommendations with interactive settings and filter. In: RecSys 2014 (2014)
5. McNee, S.M., Lam, S.K., Konstan, J.A., Riedl, J.: Interfaces for eliciting new user preferences in recommender systems. In: Brusilovsky, P., Corbett, A., Rosis, F. (eds.) UM 2003. LNCS (LNAI), vol. 2702, pp. 178–187. Springer, Heidelberg (2003). doi:10.1007/3-540-44963-9_24
6. Xiao, B., Benbasat, I.: E-commerce product recommendation agents: use, characteristics, and impact. MIS Q. **31**(1), 137–209 (2007)
7. Hijikata, Y., Kai, Y., Nishida, S.: The relation between user intervention and user satisfaction for information recommendation. In: SAC 2012, pp. 2002–2007. (2012)
8. Wasinger, R., Wallbank, J., Pizzato, L., Kay, J., Kummerfeld, B., Böhmer, M., Krüger, A.: Scrutable user models and personalised item recommendation in mobile lifestyle applications. In: Carberry, S., Weibelzahl, S., Micarelli, A., Semeraro, G. (eds.) UMAP 2013. LNCS, vol. 7899, pp. 77–88. Springer, Heidelberg (2013). doi:10.1007/978-3-642-38844-6_7
9. Knijnenburg, B.P., Reijmer, N.J., Willemsen, M.C.: Each to his own: how different users call for different interaction methods in recommender systems. In: RecSys 2011, pp. 141–148 (2011)
10. Goker, M., Thompson, C.: The adaptive place advisor: a conversational recommendation system. In: 8th German Workshop on CBR, pp. 187–198 (2000)
11. Kobsa, A., Koenemann, J., Pohl, W.: Personalized hypermedia presentation techniques for improving online customer relationships. Knowl. Eng. Rev. **16**(2), 111–155 (2001)
12. Felfernig, A., Friedrich, G., Jannach, D., Zanker, M.: An integrated environment for the development of knowledge-based recommender applications. Int. J. Electron. Commer. **11**(2), 11–34 (2006)
13. Burke, R.D., Hammond, K.J., Young, B.C.: Knowledge-based navigation of complex information spaces. In: AAAI 1996, pp. 462–468 (1996)
14. Trewin, S.: Knowledge-based recommender systems. Encyclopedia Libr. Inf. Sci. **69**, 180–200 (2000)
15. Swearingen, K., Sinha, R.: Beyond algorithms: an HCI perspective on recommender systems. In: ACM SIGIR Recommender Systems Workshop, pp. 1–11 (2001)
16. Schafer, J.B., Konstan, J.A., Riedl, J.: Meta-recommendation systems: user-controlled integration of diverse recommendations. In: CIKM 2002, pp. 43–51 (2002)
17. Schaffer, J., Höllerer, T., O'Donovan, J.: Hypothetical recommendation: a study of interactive profile manipulation behavior for recommender systems. In: FLAIRS 2015, pp. 507–512 (2015)
18. Bostandjiev, S., O'Donovan, J., Höllerer, T.: Tasteweights: a visual interactive hybrid recommender system. In: RecSys 2012, pp. 35–42 (2012)
19. Tintarev, N., Kang, B., Höllerer, T., O'Donovan, J.: Inspection mechanisms for community-based content discovery in microblogs. In: RecSys IntRS 2015 Workshop, pp. 21–28 (2015)
20. Tintarev, N., Masthoff, J.: A survey of explanations in recommender systems. In: IEEE ICDEW Data Engineering Workshop, pp. 801–810 (2007)
21. Jannach, D., Kreutler, G.: Rapid development of knowledge-based conversational recommender applications with Advisor Suite. J. Web Eng. **6**(2), 165–192 (2007)
22. Lamche, B., Adıgüzel, U., Wörndl, W.: Interactive explanations in mobile shopping recommender systems. In: RecSys IntRS 2014 Workshop, pp. 14–21 (2014)

23. Czarkowski, M., Kay, J.: A scrutable adaptive hypertext. In: Bra, P., Brusilovsky, P., Conejo, R. (eds.) AH 2002. LNCS, vol. 2347, pp. 384–387. Springer, Heidelberg (2002). doi:10.1007/3-540-47952-X_43

24. Ekstrand, M.D., Kluver, D., Harper, F.M., Konstan, J.A.: Letting users choose recommender algorithms: an experimental study. In: RecSys 2015, pp. 11–18 (2015)

25. Parra, D., Brusilovsky, P., Trattner, C.: See what you want to see: visual user-driven approach for hybrid recommendation. In: IUI 2014, pp. 235–240 (2014)

26. Harper, F.M., Xu, F., Kaur, H., Condiff, K., Chang, S., Terveen, L.G.: Putting users in control of their recommendations. In: RecSys 2015, pp. 3–10 (2015)

27. Jameson, A., Schwarzkopf, E.: Pros and cons of controllability: an empirical study. In: Bra, P., Brusilovsky, P., Conejo, R. (eds.) AH 2002. LNCS, vol. 2347, pp. 193–202. Springer, Heidelberg (2002). doi:10.1007/3-540-47952-X_21

28. Kramer, T.: The effect of measurement task transparency on preference construction and evaluations of personalized recommendations. J. Mark. Res. 44(2), 224–233 (2007)

29. Chen, L., Pu, P.: Critiquing-based recommenders: survey and emerging trends. User Model. User Adapt. Interact. 22(1–2), 125–150 (2012)

30. Groh, G., Birnkammerer, S., Köllhofer, V.: Social recommender systems. In: Ricci, F., Rokach, L., Shapira, B. (eds.) Recommender Systems for the Social Web, pp. 3–42. Springer, New York (2012)

How to Combine Visual Features with Tags to Improve Movie Recommendation Accuracy?

Yashar Deldjoo[✉], Mehdi Elahi, Paolo Cremonesi,
Farshad Bakhshandegan Moghaddam, and Andrea Luigi Edoardo Caielli

Politecnico di Milano, Milan, Italy
{yashar.deldjoo,mehdi.elahi,paolo.cremonesi}@polimi.it,
moghaddam@okit.de, andrea.caielli@mail.polimi.it
http://www.polimi.it

Abstract. Previous works have shown the effectiveness of using stylistic visual features, indicative of the movie style, in content-based movie recommendation. However, they have mainly focused on a particular recommendation scenario, *i.e.*, when a new movie is added to the catalogue and no information is available for that movie (*New Item* scenario). However, the stylistic visual features can be also used when other sources of information is available (*Existing Item* scenario).

In this work, we address the second scenario and propose a hybrid technique that exploits not only the typical content available for the movies (*e.g.*, tags), but also the stylistic visual content extracted form the movie files and fuse them by applying a fusion method called *Canonical Correlation Analysis (CCA)*. Our experiments on a large catalogue of 13 K movies have shown very promising results which indicates a considerable improvement of the recommendation quality by using a proper fusion of the stylistic visual features with other type of features.

1 Introduction

Classical approaches to multimedia recommendation are of *unimodal* nature [7,14,19,35]. Recommendations are typically generated based on two different types of item features (or attributes): metadata containing *High-Level* (or semantic) information and media entailing *Low-Level* (or stylistic) aspects.

The high-level features can be collected both from structured sources, such as databases, lexicons and ontologies, and from unstructured sources, such as reviews, news articles, item descriptions and social tags [1,2,6,10,11,27–29]. The low-level features, on the other hand, can be extracted directly from the media itself. For example, in music recommendation many acoustic features, *e.g.* rhythm and timbre, can be extracted and used to find perceptual similar tracks [3,4].

In this paper, we extend our previous works on movie recommendation [12–16], where a set of low-level visual features were used to mainly address the new item cold start scenario [17,18,34]. In such a scenario, no information is available about the new coming movies (*e.g.* user-generated movies), and the

© Springer International Publishing AG 2017
D. Bridge and H. Stuckenschmidt (Eds.): EC-Web 2016, LNBIP 278, pp. 34–45, 2017.
DOI: 10.1007/978-3-319-53676-7_3

low-level visual features are used to recommend those new movies. While this is an effective way of solving the new item problem, visual features can be also used when the other sources of information is provided (*e.g.*, tags added by users) to the movies. Accordingly, a fusion method can be used in order to combine two types of features, *i.e.* low-level stylistic features defined in our previous works [14,15] with user-generated tags into a joint representation in order to improve the quality of recommendation. Hence, we can formulate the research hypothesis as follows: *Combining the low-level visual features (extracted from movies) with tag features by a proper fusion method, can lead to more accurate recommendations, in comparison to recommendations based on these features when used in isolation.*

More particularly, we propose a *multimodal* fusion paradigm which is aimed to build a content model that exploits low-level correlation between visual-metadata modalities[1]. The method is based on *Canonical Correlation Analysis* (CCA) which belongs to a wider family of multimodal subspace learning methods known as *correlation matching* [23,30]. Unlike very few available multimodal video recommender systems todate [26,36] which treat the fusion problem as a basic linear modeling problem without studying the underlying spanned feature spaces, the proposed method learns the correlation between modalities and maximize the pairwise correlation.

The main contributions of this work are listed below:

- we propose a novel technique that combines a set of automatically extracted stylistic visual features with other source of information in order to improve the quality of recommendation.
- we employ a data fusion method called *Canonical Correlation Analysis (CCA)* [20,21] that unlike traditional fusion methods, which do not exploit the relationship between two set of features coming from two different sources, achieves this by maximizing the pairwise correlation between two sets.
- we evaluate our proposed technique with a large dataset with more than 13 K movies that has been thoroughly analyzed in order to extract the stylistic visual features[2].

The rest of the paper is organized as follows. Section 2 briefly reviews the related work. Section 3 discusses the proposed method by presenting a description of the visual features and introducing a mathematical model for the recommendation problem and the proposed fusion method. Section 4 presents the evaluation methodology. Results and discussions are presented in Sect. 5. Finally, in Sect. 6 we present the conclusion and future work.

---

[1] Note that though textual in nature, we treat metadata as a separate modality which is added to a video by a community-user (tag) or an expert (genre). Refer to Table 1 for further illustration.

[2] The dataset is called *Mise-en-scene Dataset* and it is publicly available through the following link: http://recsys.deib.polimi.it.

2 Related Work

Up to the present, the exploitation of low-level features have been marginally explored in the community of recommender systems. This is while such features have been extensively studied in other fields such as computer vision and content-based video retrieval [24,32]. Although for different objectives, these communities share with the community of recommender systems, the research problems of defining the "best" representation of video content and of classifying videos according to features of different nature. Hence they offer results and insights that are of interest also in the movie recommender systems context.

The works presented in [5,24] provide comprehensive surveys on the relevant state of the art related to video content analysis and classification, and discuss a large body of low-level features (visual, auditory or textual) that can be considered for these purposes. In [32] Rasheed et al. proposes a practical movie genre classification scheme based on computable visual cues. [31] discusses a similar approach by considering also the audio features. Finally, in [37] Zhou et al. proposes a framework for automatic classification, using a temporally-structured features, based on the intermediate level of scene representation.

While the scenario of using the low-level features has been interesting for the goal of video retrieval, this paper addresses a different scenario, *i.e.*, when the the low-level features features are used in a recommender system to effectively generate relevant recommendations for users.

3 Method Descriptions

In this section, we present the proposed method.

3.1 Visual Features

Multimedia content in a video can be classified into three hierarchical levels:

- **Level 1 "High-level (semantic) features":** At this level, we have *semantic* features that deal with the concepts and events happening in a video. For example, the plot of the movie "The Good, the Bad and the Ugly", which revolves around three gunslingers competing to find a buried cache of gold during the American Civil War.
- **Level 2 "Mid-level (syntactic) features":** At the intermediate level we have *syntactic features* that deal with what objects exist in a video and their interactions. As an example, in the same movie there are Clint Eastwood, Lee Van Cleef, Eli Wallach, plus several horses and guns.
- **Level 3 "Low-level (stylistic) features":** At the lowest level we have *stylistic features* which define the Mise-en-Scene characteristics of the movie, *i.e.*, the design aspects that characterize aesthetic and style of a movie. As an example, in the same movie predominant colors are yellow and brown, and camera shots use extreme close-up on actors' eyes.

The examples above are presented for the visual modality as it forms the focus of our recent works [13–15]. In order to allow a fair comparison between different modalities, we present the hierarchical comparison of multimedia content across different modalities [7] in Table 1. Note that while the visual, aural and textual modalities are elements of the multimedia data itself, metadata is *added* to the movie after production.

Table 1. Hierarchical and modality-wise classification of multimedia features

Level	Visual	Aural	Text	Metadata
High	events, concepts	events, concepts	semantic similarity	summary, tag
Mid	objects/people, objects' interaction	objects/people, source	sentences, keywords	genre, tag, cast
Low	motion, color, shape, lighting	timbre, pitch spectral frequency	nouns, verbs, adjectives	genre, tag

Recommender systems in the movie domain typically use high-level or mid-level features such as genre or tag which appears in the form of metadata [19, 25, 35]. These feature usually cover a wide range in the hierarchical classification of content, for example tags most often contain words about events and incidents (high-level), people and places (mid-level) while visual features extracted and studied in our previous works (presented in Table 2) cover the low-level aspects. By properly combining the high-level metadata and low-level visual features we aim to maximize the informativeness of the joint feature representation.

3.2 Multimedia Recommendation Problem

A multimedia document D (*e.g.* a video) can be represented with the quadruple

$$D = (d_V, d_A, d_T, d_M)$$

in which d_V, d_A, d_T are the *visual*, *aural* and *textual* documents constituting a multimedia document and d_M is the *metadata* added to the multimedia document by a human (*e.g.* tag, genre or year of production). In a similar manner, a user's profile U can be projected over each of the above modalities and be symbolically represented as

$$U = (u_V, u_A, u_T, u_M)$$

The multimedia components are represented as vectors in features spaces $\mathbb{R}^{|V|}$, $\mathbb{R}^{|A|}$, $\mathbb{R}^{|T|}$ and $\mathbb{R}^{|M|}$. For instance, $f_V = \{f_1, ..., f_{|V|}\}^T \in \mathbb{R}^{|V|}$ is the feature vector representing the visual component. The relevancy between the target user

Table 2. The list of low-level stylistic features representative of movie style presented in our previous works [14, 16]

Features	Equation	Description
Camera motion	$\overline{L}_{sh} = \frac{n_f}{n_{sh}}$	Camera shot is used as the representative measure of camera movement. A shot is a single camera action. The Average shot length $\overline{L}_{sh}$ and number of shots n_{sh} are used as two distinctive features
Color variance	$\rho = \begin{pmatrix} \sigma_L^2 & \sigma_{Lu}^2 & \sigma_{Lv}^2 \\ \sigma_{Lu}^2 & \sigma_u^2 & \sigma_{uv}^2 \\ \sigma_{Lv}^2 & \sigma_{uv}^2 & \sigma_v^2 \end{pmatrix}$	For each keyframe in the Luv colorspace the covariance matrix ρ is computed where $\sigma_L, \sigma_u, \sigma_v, \sigma_{Lu}, \sigma_{Lv}, \sigma_{uv}$ are the standard deviation over three channels L, u, v and their mutual covariance. $\Sigma = det(\rho)$ is the measure for color variance. The mean and std of Σ over keyframes are used as the representative features of color variance
Object motion	$\nabla I(x, t).v + I_t(x, t) = 0$	Object motion is calculated based on optical flow estimation. which provides a robust estimate of the object motion in video frames based on pixel velocities. The mean and std of of pixels motion is calculated on each frames and averaged across all video as the two representative features for object motion
Lighting key	$\xi = \mu.\sigma$	After transforming pixel to HSV colorspace the mean μ and std σ of the value component which corresponds to the brightness is computed. ξ which is the multiplication of two is computed and averaged across keyframes as the measure of average lighting key in a video

profile U and the item profile D is of interest for recommenders and is denoted with $\mathcal{R}(U, D)$. In this work, we will focus our attention to the visual features defined in Table 2 and the rich metadata (tag).

Given a user profile U either directly provided by her (direct profile) or evaluated by the system (indirect profile) and a database of videos $\mathcal{D} = \{D_1, D_2, ..., D_{|D|}\}$, the task of video recommendation is to seek the video D_i

that satisfies

$$D_i^* = \arg \max_{D_i \in \mathcal{D}} \mathcal{R}(U, D_i) \tag{1}$$

where $\mathcal{R}(U, D)$ can be computed as

$$\mathcal{R}(U, D_i) = R(F(u_V, u_M), F(d_V, d_M)) \tag{2}$$

where F is a function whose role is to combine different modalities into a joint representation. This function is known by *inter-modal fusion function* in multimedia information retrieval (MMIR). It belongs to the family of fusion methods known as *early fusion* methods which integrates unimodal features before passing them to a recommender. The effectiveness of early fusion methods has been proven in couple of multimedia retrieval papers [30,33].

3.3 Fusion Method

The fusion method aims to combine information obtained from two sources of features: (1) *LL Features*: stylistic visual features extracted by our system and (2) *HL features*: the tag features. We employ a novel fusion method known as *Canonical Correlation Analysis* (CCA) [20,21,30] for fusing two sources of features. CCA is popular method in multi-data processing and is mainly used to analyse the relationships between two sets of features originated from different sources of information.

Given two set of features $X \in R^{p \times n}$ and $Y \in R^{q \times n}$, where p and q are the dimension of features extracted from the n items, let $S_{xx} \in R^{p \times p}$ and $S_{yy} \in R^{q \times q}$ be the *between-set* and $S_{xy} \in R^{p \times q}$ be the *within-set* covariance matrix. Also let us define $S \in R^{(p+q) \times (p+q)}$ to be the *overall covariance matrix* - a complete matrix which contains information about association between pairs of features-represented as following

$$S = \begin{pmatrix} \text{cov}(x) & \text{cov}(x,y) \\ \text{cov}(y,x) & \text{cov}(y) \end{pmatrix} = \begin{pmatrix} S_{xx} & S_{xy} \\ S_{yx} & S_{yy} \end{pmatrix} \tag{3}$$

then CCA aims to find a linear transformation $X^* = W_x^T.X$ and $Y^* = W_y^T.Y$ that maximizes the pair-wise correlation across two feature set as given by Eq. 4. The latter will ensure the relationship between two set of features will follow a consistent pattern. This would lead to creation of discriminative and informative fused feature vector

$$\arg \max_{W_x, W_y} \text{corr}(X^*, Y^*) = \frac{\text{cov}(X^*, Y^*)}{\text{var}(X^*).\text{var}(Y^*)} \tag{4}$$

where $\text{cov}(X^*, Y^*) = W_x^T S_{xy} W_y$ and for variances we have that $\text{var}(X^*) = W_x^T S_{xx} W_x$ and $\text{var}(Y^*) = W_y^T S_{yy} W_y$. We adopt the maximization procedure described in [20,21] and solving the eigenvalue equation

$$\begin{cases} S_{xx}^{-1} S_{xy} S_{yy}^{-1} S_{yx} \hat{W}_x = \Lambda^2 \hat{W}_x \\ S_{yy}^{-1} S_{yx} S_{xx}^{-1} S_{xy} \hat{W}_y = \Lambda^2 \hat{W}_y \end{cases} \tag{5}$$

where $W_x, W_y \in R^{p \times d}$ are the eigenvectors and Λ^2 is the diagonal matrix of eigenvalues or squares of the *canonical correlations*. Finally, $d = rank(S_{xy}) \leq \min(n, p, q)$ is the number of non-zero eigenvalues in each equation. After calculating $X^*, Y^* \in R^{d \times n}$, feature-level fusion can be performed in two manners: (1) concatenation (2) summation of the transformed features:

$$Z^{ccat} = \begin{pmatrix} X^* \\ Y^* \end{pmatrix} = \begin{pmatrix} W_x^T.X \\ W_y^T.Y \end{pmatrix} \tag{6}$$

and

$$Z^{sum} = X^* + Y^* = W_x^T.X + W_y^T.Y \tag{7}$$

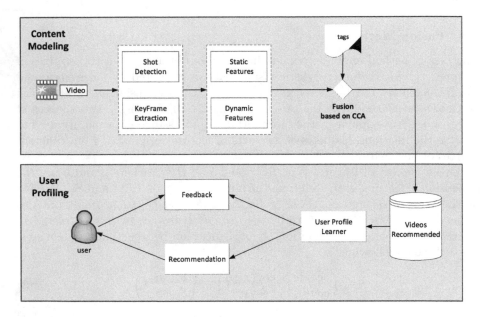

Fig. 1. Illustration of the proposed video recommender system based on stylistic low-level visual feature and user-generated tag using a fusion method based on CCA

Figure 1 illustrates the building block of the developed video recommender system. Color variance and lighting key are the extracted static features and camera and object motion are the dynamic features.

3.4   Recommendation Algorithm

To generate recommendations, we adopted a classical "k-nearest neighbor" content-based algorithm. Given a set of users U and a catalogue of items I, a set of preference scores r_{ui} has been collected. Moreover, each item $i \in I$ is associated to its feature vector f_i. For each couple of items i and j, a similarity score s_{ij} is computed using *cosine similarity* as follows

$$s_{ij} = \frac{f_i^T f_j}{\|f_i\| \|f_j\|} \tag{8}$$

For each item i the set of its nearest neighbors NN_i is built, $|NN_i| < K$. Then, for each user $u \in U$, the predicted preference score $\hat{r_{ui}}$ for an unseen item i is computed as follows

$$\hat{r_{ui}} = \frac{\sum_{j \in NN_i, r_{uj} > 0} r_{uj} s_{ij}}{\sum_{j \in NN_i, r_{uj} > 0} s_{ij}} \tag{9}$$

4 Evaluation Methodology

4.1 Dataset

We have used the latest version of Movielens dataset [22] which contains 22'884'377 ratings and 586'994 tags provided by 247'753 users to 34'208 movies (sparsity 99.72%). For each movie in Movielens dataset, the title has been automatically queried in YouTube to search for the trailer. If the trailer is available, it has been downloaded. We have found the trailers for 13'373 movies.

Low-level features have been automatically extracted from trailers. We have used trailers and not full videos in order to have a scalable recommender system. Previous works have shown that low-level features extracted from trailers of movies are equivalent to the low-level features extracted from full-length videos, both in terms of feature vectors and quality of recommendations [14].

We have used Latent Semantic Analysis (LSA) to better exploit the implicit structure in the association between tags and items. The technique consists in decomposing the tag-item matrix into a set of orthogonal factors whose linear combination approximates the original matrix [8].

4.2 Methodology

We have evaluated the Top-N recommendation quality by adopting a procedure similar to the one described in [9].

- We split the dataset into two random subsets. One of the subsets contains 80% of the ratings and it is used for training the system (train set) and the other one contains 20% of the rating and it is used for evaluation (test set).
- For each relevant item i rated by user u in test set, we form a list containing the item i and all the items not rated by the user u, which we assume to be irrelevant to her. Then, we formed a top-N recommendation list by picking the top N ranked items from the list. Being r the rank of i, we have a *hit* if $r < N$, otherwise we have a *miss*. Hence, if a user u has N_u relevant items, the precision and recall in its recommendation list of size N is computed.
- We measure the quality of the recommendation in terms of recall, precision and mean average precision (MAP) for different cutoff values $N = 5, 10, 20$.

5 Results

Table 3 represents the results that we have obtained from the conducted experiments. As it can be seen, both methods for fusion of LL visual features with TagLSA features, outperform either of using the TagLSA and LL visual features, with respect to the all considered evaluation metrics.

In terms of recall, fusion of LL visual with TagLSA, based on concatenation of these features (ccat), obtains the score of 0.0115, 0.0166, and 0.0209 for recommendation at 5, 10, and 20, respectively. The alternative fusion method based on summation (sum), also scores better than the other baselines, $i.e.$, LL visual features and TagLSA, with the recall values of 0.0055, 0.0085, and 0.0112 for different recommendation sizes (cutoff values). These values are 0.0038, 0.0046, and 0.0053 for recommendation by using LL visual features and 0.0028, 0.0049, and 0.0068 for recommendation by using TagLSA. These scores indicate that recommendation based on fusion of LL visual features and TagLSA features is considerably better than recommendation based on these content features individually.

In terms of precision, again, the best results are obtained by fusion of LL visual features with TagLSA features based on concatenation with scores of 0.0140, 0.0115, and 0.0079 for recommendation at 5, 10, and 20, respectively. The alternative fusion method (sum) obtains precision scores of 0.0081, 0.0069, and 0.0048 which is better than the other two individual baselines. Indeed, LL visual features archives precision scores of 0.0051, 0.0037, and 0.0023, while the TagLSA achieves precision scores of 0.0045, 0.0041, and 0.0031. These results also indicates the superior quality of the recommendation based on fusion of LL visual features and TagLSA in comparison to recommendation each of these set of features.

Similar results have been obtained in terms of MAP metric. Hence, fusion method based on concatenation (ccat) performs the best in comparison to the other baselines, by obtaining the MAP scores of 0.0091, 0.0080, and 0.0076 for recommendation at 5, 10, and 20. The MAP scores are 0.0045, 0.0038, 0.0035 for fusion of features based on summation (sum), 0.0035, 0.0028, 0.0026 for LL visual features, and 0.0025, 0.0021, 0.0019 for TagLSA. Accordingly, the fusion of the LL visual features and TagLSA presents excellent performance in terms of MAP metric.

Table 3. Quality of recommendation w.r.t Recall, Precision and MAP when using low-level visual features and high-level metadata features in isolation compared with fused features using our proposed method based on Canonical Correlation Analysis.

Features	Fusion method	Recall			Precision			MAP		
		@5	@10	@20	@5	@10	@20	@5	@10	@20
TagLSA	-	0.0028	0.0049	0.0068	0.0045	0.0041	0.0031	0.0025	0.0021	0.0019
LL	-	0.0038	0.0046	0.0053	0.0051	0.0037	0.0023	0.0035	0.0028	0.0026
LL + TagLSA	CCA-Sum	0.0055	0.0085	0.0112	0.0081	0.0069	0.0048	0.0045	0.0038	0.0035
LL + TagLSA	CCA-Ccat	**0.0115**	**0.0166**	**0.0209**	**0.0140**	**0.0115**	**0.0079**	**0.0091**	**0.0080**	**0.0076**

Overall, the results validates our hypothesis and shows that combining the Low-Level visual features (LL visual) extracted from movies with tag content, by adopting a proper fusion method, can lead to significant improvement on the quality of recommendations. This is an promising outcome and shows the great potential of exploiting LL visual features together with other sources of content information such as tags in generation of relevant personalised recommendation in multimedia domain.

6   Conclusion and Future Work

In this paper, we propose the fusion of visual features extracted from the movie files with other types of content (*i.e.*, tags), in order to improve the quality of the recommendation. In the previous works, the visual features are used mainly to solve cold start problem, *i.e.*, when a new movie is added to the catalogue and no information is available for that movie. In this work, however, we use the stylistic visual features in combination with other sources of information. Hence, our research hypothesis is that a proper fusion of the visual features of movies may have led to a higher accuracy of movie recommendation, *w.r.t.* using these set of features individually.

Based on the experiments, we conducted on a large dataset of 13 K movies, we successfully verified the hypothesis and shown that the recommendation accuracy is considerably improved when the (low-level) visual features are combined with user-generated tags.

In future, we would consider fusion of additional sources of information, such as, audio features, in order to farther improve the quality of the content based recommendation system. Moreover, we will investigate the effect of different feature aggregation methods on the quality of the extracted information and on the quality of the generated recommendations.

References

1. Aggarwal, C.C.: Content-based recommender systems. In: Recommender Systems, pp. 139–166. Springer, Heidelberg (2016)
2. Aggarwal, C.C.: Recommender Systems: The Textbook. Springer, Heidelberg (2016)
3. Bogdanov, D., Herrera, P.: How much metadata do we need in music recommendation? a subjective evaluation using preference sets. In: ISMIR, pp. 97–102 (2011)
4. Bogdanov, D., Serrà, J., Wack, N., Herrera, P., Serra, X.: Unifying low-level and high-level music similarity measures. IEEE Trans. Multimedia **13**(4), 687–701 (2011)
5. Brezeale, D., Cook, D.J.: Automatic video classification: A survey of the literature. IEEE Trans. Syst. Man Cybern. Part C: Appl. Rev. **38**(3), 416–430 (2008)
6. Cantador, I., Szomszor, M., Alani, H., Fernández, M., Castells, P.: Enriching ontological user profiles with tagging history for multi-domain recommendations (2008)
7. Celma, O.: Music recommendation. In: Music Recommendation and Discovery, pp. 43–85. Springer, Heidelberg (2010)

8. Cremonesi, P., Garzotto, F., Negro, S., Papadopoulos, A.V., Turrin, R.: Looking for "Good" recommendations: a comparative evaluation of recommender systems. In: Campos, P., Graham, N., Jorge, J., Nunes, N., Palanque, P., Winckler, M. (eds.) INTERACT 2011. LNCS, vol. 6948, pp. 152–168. Springer, Heidelberg (2011). doi:10.1007/978-3-642-23765-2_11

9. Cremonesi, P., Koren, Y., Turrin, R.: Performance of recommender algorithms on top-n recommendation tasks. In: Proceedings of the ACM Conference on Recommender Systems, RecSys 2010, Barcelona, Spain, pp. 39–46, 26–30 September 2010

10. de Gemmis, M., Lops, P., Musto, C., Narducci, F., Semeraro, G.: Semantics-aware content-based recommender systems. In: Ricci, F., Rokach, L., Shapira, B. (eds.) Recommender Systems Handbook, pp. 119–159. Springer, Heidelberg (2015)

11. Degemmis, M., Lops, P., Semeraro, G.: A content-collaborative recommender that exploits wordnet-based user profiles for neighborhood formation. User Model. User-Adap. Inter. 17(3), 217–255 (2007)

12. Deldjoo, Y., Elahi, M., Cremonesi, P.: Using visual features and latent factors for movie recommendation. In: Workshop on New Trends in Content-Based Recommender Systems (CBRecSys), in Conjugation with ACM Recommender Systems Conference (RecSys) (2016)

13. Deldjoo, Y., Elahi, M., Cremonesi, P., Garzotto, F., Piazzolla, P.: Recommending movies based on mise-en-scene design. In: Proceedings of the 2016 CHI Conference Extended Abstracts on Human Factors in Computing Systems, pp. 1540–1547. ACM (2016)

14. Deldjoo, Y., Elahi, M., Cremonesi, P., Garzotto, F., Piazzolla, P., Quadrana, M.: Content-based video recommendation system based on stylistic visual features. J. Data Semant. 5(2), 99–113 (2016)

15. Deldjoo, Y., Elahi, M., Quadrana, M., Cremonesi, P.: Toward building a content-based video recommendation system based on low-level features. In: Stuckenschmidt, H., Jannach, D. (eds.) EC-Web 2015. LNBIP, vol. 239, pp. 45–56. Springer, Heidelberg (2015). doi:10.1007/978-3-319-27729-5_4

16. Deldjoo, Y., Elahi, M., Quadrana, M., Cremonesi, P., Garzotto, F.: Toward effective movie recommendations based on mise-en-scène film styles. In: Proceedings of the 11th Biannual Conference on Italian SIGCHI Chapter, pp. 162–165. ACM (2015)

17. Elahi, M., Ricci, F., Rubens, N.: A survey of active learning in collaborative filtering recommender systems. Comput. Sci. Rev. 20, 29–50 (2016)

18. Fernández-Tobías, I., Braunhofer, M., Elahi, M., Ricci, F., Cantador, I.: Alleviating the new user problem in collaborative filtering by exploiting personality information. User Model. User-Adap. Inter. 26(2), 221–255 (2016)

19. Guy, I., Zwerdling, N., Ronen, I., Carmel, D., Uziel, E.: Social media recommendation based on people and tags. In: Proceedings of the 33rd International ACM SIGIR Conference on Research and Development in Information Retrieval, pp. 194–201. ACM (2010)

20. Haghighat, M., Abdel-Mottaleb, M., Alhalabi, W.: Fully automatic face normalization and single sample face recognition in unconstrained environments. Expert Syst. Appl. 47, 23–34 (2016)

21. Hardoon, D.R., Szedmak, S., Shawe-Taylor, J.: Canonical correlation analysis: an overview with application to learning methods. Neural Comput. 16(12), 2639–2664 (2004)

22. Harper, F.M., Konstan, J.A.: The movielens datasets: History and context. ACM Trans. Interact. Intell. Syst. (TiiS) 5(4), 19 (2015)

23. Hotelling, H.: Relations between two sets of variates. Biometrika 28(3/4), 321–377 (1936)

24. Hu, W., Xie, N., Li, L., Zeng, X., Maybank, S.: A survey on visual content-based video indexing and retrieval. IEEE Trans. Syst. Man Cybern. Part C: Appl. Rev. **41**(6), 797–819 (2011)
25. Li, X., Guo, L., Zhao, Y.E.: Tag-based social interest discovery. In: Proceedings of the 17th International Conference on World Wide Web, pp. 675–684. ACM (2008)
26. Mei, T., Yang, B., Hua, X.-S., Li, S.: Contextual video recommendation by multi-modal relevance and user feedback. ACM Trans. Inf. Syst. (TOIS) **29**(2), 10 (2011)
27. Mooney, R.J., Roy, L.: Content-based book recommending using learning for text categorization. In: Proceedings of the Fifth ACM Conference on Digital Libraries, pp. 195–204. ACM (2000)
28. Nasery, M., Braunhofer, M., Ricci, F.: Recommendations with optimal combination of feature-based and item-based preferences. In: Proceedings of the 2016 Conference on User Modeling Adaptation and Personalization, pp. 269–273. ACM (2016)
29. Nasery, M., Elahi, M., Cremonesi, P.: Polimovie: a feature-based dataset for recommender systems. In: ACM RecSys Workshop on Crowdsourcing and Human Computation for Recommender Systems (CrawdRec), vol. 3, pp. 25–30. ACM (2015)
30. Pereira, J.C., Coviello, E., Doyle, G., Rasiwasia, N., Lanckriet, G.R., Levy, R., Vasconcelos, N.: On the role of correlation and abstraction in cross-modal multimedia retrieval. IEEE Trans. Pattern Anal. Mach. Intell. **36**(3), 521–535 (2014)
31. Rasheed, Z., Shah, M.: Video categorization using semantics and semiotics. In: Rosenfeld, A., Doermann, D., DeMenthon, D. (eds.) Video Mining. The Springer International Series in Video Computing, vol. 6, pp. 185–217. Springer, Heidelberg (2003)
32. Rasheed, Z., Sheikh, Y., Shah, M.: On the use of computable features for film classification. IEEE Trans. Circuits Syst. Video Technol. **15**(1), 52–64 (2005)
33. Rasiwasia, N., Costa Pereira, J., Coviello, E., Doyle, G., Lanckriet, G.R., Levy, R., Vasconcelos, N.: A new approach to cross-modal multimedia retrieval. In: Proceedings of the 18th ACM International Conference on Multimedia, pp. 251–260. ACM (2010)
34. Rubens, N., Elahi, M., Sugiyama, M., Kaplan, D.: Active learning in recommender systems. In: Ricci, F., Rokach, L., Shapira, B. (eds.) Recommender Systems Handbook, pp. 809–846. Springer, Heidelberg (2015)
35. Szomszor, M., Cattuto, C., Alani, H., O'Hara, K., Baldassarri, A., Loreto, V., Servedio, V.D.: Folksonomies, the semantic web, and movie recommendation (2007)
36. Yang, B., Mei, T., Hua, X.-S., Yang, L., Yang, S.-Q., Li, M.: Online video recommendation based on multimodal fusion and relevance feedback. In: Proceedings of the 6th ACM International Conference on Image and Video Retrieval, pp. 73–80. ACM, (2007)
37. Zhou, H., Hermans, T., Karandikar, A.V., Rehg, J.M.: Movie genre classification via scene categorization. In: Proceedings of the International Conference on Multimedia, pp. 747–750. ACM (2010)

Explorative Analysis of Recommendations Through Interactive Visualization

Christian Richthammer$^{(\boxtimes)}$ and Günther Pernul

Department of Information Systems, University of Regensburg, Regensburg, Germany
{christian.richthammer,guenther.pernul}@wiwi.uni-regensburg.de

Abstract. Even though today's recommender algorithms are highly sophisticated, they can hardly take into account the users' situational needs. An obvious way to address this is to initially inquire the users' momentary preferences, but the users' inability to accurately state them upfront may lead to the loss of several good alternatives. Hence, this paper suggests to generate the recommendations without such additional input data from the users and let them interactively explore the recommended items on their own. To support this explorative analysis, a novel visualization tool based on treemaps is developed. The analysis of the prototype demonstrates that the interactive treemap visualization facilitates the users' comprehension of the big picture of available alternatives and the reasoning behind the recommendations. This helps the users get clear about their situational needs, inspect the most relevant recommendations in detail, and finally arrive at informed decisions.

Keywords: Recommender systems · Interactive visualization · Search space · Explorative analysis

1 Introduction

Nowadays, the ever-increasing amounts of alternatives and information overwhelm people in their decision making processes. Recommender systems, which emerged in the mid 1990s [1–3] and formed a research field of their own since then, constitute an important approach to address this phenomenon of information overload [4,5]. Their main aim is to inform users about items they will probably like and have previously not been aware of. This is also referred to as the recommendation problem [1,4]. For many years, the dominating topic in the recommender systems community has been the development of prediction algorithms along with the evaluation of their recommendation accuracy [5,6], experiencing its peak with the Netflix Prize [7]. As the development of new algorithms and the optimization of existing ones often result in only marginal improvements by now [8,9], a shift of research interests to the ultimate user experience can be observed [10]. Researchers have realized that algorithms may be the backbone of any good recommender system but that the users' evaluation of the system and its recommendations is crucial [11,12].

© Springer International Publishing AG 2017
D. Bridge and H. Stuckenschmidt (Eds.): EC-Web 2016, LNBIP 278, pp. 46–57, 2017.
DOI: 10.1007/978-3-319-53676-7_4

Traditional recommender algorithms can, even in their sophisticatedly optimized form, only produce a set of recommendations that may generally be of interest to the users. Considering that the users may have very special needs (notably deviating from their preferences determined according to their past rating behaviors) in a particular situation, the top recommendation does not necessarily have to be the best choice in that situation. An obvious way to address this is to initially inquire the users' situational needs as done in knowledge-based recommender systems [13]. The users' input can then be used for filtering or weighting processes. However, the users are unlikely to be able to accurately state all of their preferences upfront, especially without having seen any suggestions and if there are any suitable alternatives fulfilling their criteria at all [9,14]. Thus, filtering may lead to the loss of several alternatives that are worth considering but lie just outside the defined filtering criteria.

In contrast to the aforementioned initial inquiry of the users' situational needs, this paper suggests to generate the set of recommendations without such additional input data from the users and let them interactively explore the recommended items on their own with their situational needs in mind. We aim to support this explorative analysis by relying on interactive visualization, a concept that has been ascribed a lot of uncovered potential in the field of recommender systems [8,15]. In particular, we develop a novel visualization tool based on treemaps. Treemaps arrange the recommendations in a way that provides a structured overview of large parts of the search space (i.e. the set of all available alternatives), which facilitates the users' comprehension of the big picture of options fitting their needs [16]. On the one hand, educating the users about the search space helps them understand the reasoning behind the recommendations [17]. On the other hand, it serves the users as a basis to get clear about their situational needs, reduce the number of displayed recommendations to the top ones, and inspect these in detail.

The remainder of the paper is structured according to the guidelines for conducting design science research by Hevner et al. [18]. In particular, it follows the six activities of the design science research methodology introduced by Peffers et al. [19]. In this section, the first activity (*problem identification and motivation*) is covered by introducing the problem context and expounding the motivation for the explorative analysis of recommendations. In Sect. 2, related work on visualizations for recommender systems are discussed and the objectives of our solution are pointed out as demanded by the second activity (*define the objectives for a solution*). The third activity (*design and development*) is split into two steps: presenting the conceptual design of the treemap visualization in Sect. 3, and describing its prototypical implementation in Sect. 4. Section 5 is dedicated to the demonstration of our solution and the evaluation of our research objectives, thus covering the fourth (*demonstration*) and fifth activity (*evaluation*). The discussion of aspects for future work in Sect. 6 concludes the paper. Publishing our findings covers the sixth activity (*communication*).

2   Related Work and Research Objectives

The research interest in visualizations of recommendations that are more sophisticated than simple lists has been growing steadily since the end of the last
decade. One of the first notable proposals in this area is PeerChooser [20], whose
graph-based interface is supposed to provide a visual explanation of the collaborative filtering process and to enable the users to interactively manipulate their
neighborhood. Also using a graph-based interface, SmallWorlds [21] allows to
express item preference profiles with which personalized item recommendations
are generated in a transparent manner. This enables the users both to explore
their relationships to their peers and to discover new relevant items. Exploring relationships to peers is also the motivation behind SFViz [22]. Here, friend
recommendations in social networks are visualized using a radial space-filling
technique as well as a circle layout with edge bundling. Residing in a similar application area, TasteWeights [23] depicts item predictions based on social
web resources (e.g. Wikipedia, Facebook, and Twitter). Its main characteristic is
that the users are able to interactively adjust the weights of the different context
sources and the corresponding items. Assigning weights to different recommendation strategies is an important feature in SetFusion [24], too.

In recent years, a number of researchers have used map-based interfaces to support the users in understanding and exploring recommendations. In particular,
Gansner et al. [25] use geographic maps for TV show recommendations, Kagie et al.
[26] use two-dimensional maps for product recommendations in e-commerce, and
Verbert et al. [27] use clustermaps for talk recommendations at scientific conferences. Specifically concentrating on treemaps, Tintarev and Masthoff [16,28] state
that this kind of visualization has not been used for recommender systems even
though it may be a valuable choice. Based on our literature search, we can verify
this claim and confirm that it still holds. The only proposal going into this direction
is the treemap-like component presented by Katarya et al. [29], which is included
in their interface containing several visual explanation styles for movie recommendations. We call it "treemap-like" because we are missing two distinctive properties of treemaps (cf. Sect. 3). First, the visualization does not use the available
space to its full extent. And second, it does not display any hierarchical structures.
In this work, we employ the treemap visualization in its traditional form because
we believe these two properties to be important for depicting the search space as
completely as possible. Furthermore, the focus of Katarya et al. [29] differs from
ours. They concentrate on different ways to visualize explanations for recommendations, whereas explanations are rather an add-on in our proposal. We believe that
a good visualization can improve user satisfaction by itself, which is why we lay our
primary focus on the detailed examination of the treemap.

The analysis of related work shows that despite the generally increasing
interest in interactive visualizations for recommender systems, the employment
of treemaps has received comparatively little attention so far. Since this kind
of visualization has proven to be effective in connection with non-personalized
news recommendations [30], we want to explore this promising concept in its
traditional form for personalized recommendations as well. The overall goal of

our treemap visualization is to support the users' explorative analysis of recommendations. In particular, we want to achieve this by pursuing the following
objectives:

1. Facilitate the users' comprehension of the recommendations by educating
   them about the search space.
2. Enable the users to interactively explore the recommendations with their
   situational needs in mind.
3. Increase the transparency of the recommender system by helping the users
   understand the accomplishment of the recommendations.

3 Conceptual Design

The treemap (first designed in the early 1990s [31,32]) is a space-constrained
visualization that uses nested rectangles to display large amounts of monohierarchical, tree-structured data with great efficiency and high readability [33].
The sizes of the rectangles are calculated by the underlying tiling algorithm, for
which several popular alternatives with different characteristics are available.
Moreover, it is possible to use different colors and different degrees of opacity for
the rectangles, which enables displaying additional attributes. Ways to display
even more attributes include different font sizes, font styles, and borders of the
rectangles. However, we limit the employment of dimensions to the following
four because we do not want to clutter our treemap visualization.

Hierarchy. As its name suggests and as already pointed out, the treemap is
   used to visualize tree-structured data. In contrast to the size dimension, the
   overarching hierarchy dimension is not used for quantitative data but for
   distinct categories (e.g. different states when visualizing voting data of US
   presidential elections).
Size. The size dimension determines the size of the single rectangles. It should
   represent a quantifiable attribute so that the sizes of the rectangles at a lower
   hierarchical level can be summed up to the size of the superordinate rectangle.
   The largest rectangle is arranged in the top left corner.
Color. The color dimension determines the color of the single rectangles. It is most
   qualified to depict a measure of performance, change, or development [34]. However, it can also be used to support the visualization of the hierarchy dimension by assigning different colors to the different elements on a certain hierarchy
   level.
Opacity. The opacity dimension can be used in addition to the color dimension
   and in order to reduce the degree of visibility of items that are less relevant
   according to a specified criterion. In a news recommender system, for example,
   older stories may be less interesting than more recent stories regardless of their
   overall predicted importance.

Figure 1 provides a conceptual overview of our visualization component
including the dimensions introduced before. As can be seen in the bottom part of

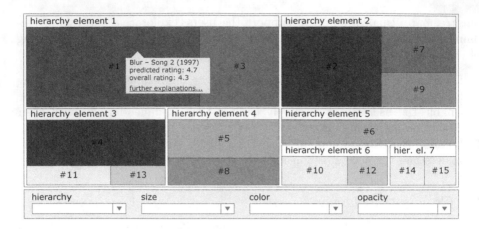

Fig. 1. Conceptual overview of the treemap visualization component.

the figure, we want the users to be able to change the assignment of the available attributes to the dimensions in order to give them full control over the visualization. In addition, the users are provided with comprehensive information on the items in the form of a tooltip. This includes general metadata (e.g. title, artist, and release year in a music recommender) but also recommendation-specific data such as predicted rating, overall rating, and further explanations. When clicking on a treemap element, the corresponding tooltip is expanded (cf. Fig. 4).

Although this proposal's focus is on the pure visualization, we include explanations as an add-on to further increase the user acceptance of recommendations [35]. The positive effect on the user acceptance mainly stems from the increase in transparency that follows from the employment of explanations. When providing explanations, recommender systems are no black boxes anymore. Instead, they feel like the digital analogue to the transparent social process of recommendation-making in the real world [36]. Specifically, we add a histogram (cf. Fig. 4) showing the ratings of other users to each item because this easy-to-understand form of explanation performed especially well in Herlocker et al.'s [37] pioneering study on the persuasiveness of explanations. In contrast to this, we also provide detailed information on the calculations and intermediate values of the algorithms such as the similarity values of the nearest neighbors. Because of the complexity of the latter form of explanation, the corresponding information is collapsed by default but can be expanded by interested users (cf. Fig. 4).

4   Implementation

In this section, we demonstrate the feasibility of the conceptual design by implementing it in a software prototype. We base the implementation on the popular MovieLens 100K dataset (100,000 ratings from 943 users on 1,682 movies) [38] because its list of entities and attributes perfectly fits our conceptual design:

– rating: user ID, movie ID, rating, timestamp
– user: user ID, age, gender, occupation, zip code
– movie: movie ID, title, release date, IMDb URL, genre

The release date, the genre, the popularity (number of ratings), and the average rating of the movies can be employed as attributes for the treemap dimensions introduced before. The age, the gender, the occupation, and the zip code of the users can be employed for the explanation of user-based collaborative filtering algorithms. The IMDb URL is needed to provide additional data on the movies. The list of attributes is completed by the predicted rating of the movies, which is determined by the recommender algorithm and used as another attribute for the treemap dimensions.

Figure 2 provides an overview of the architecture underlying our prototype. In the backend, we use the LAMP solution stack, which consists of a Linux operating system, an Apache web server, a MySQL database management system, and the PHP programming language. The frontend of the prototype is realized using HTML5, CSS, and the JavaScript framework jQuery. Thus, it is platform-independent and can easily be accessed through a web browser.

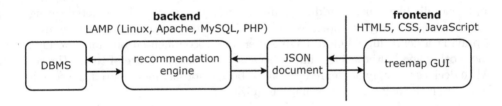

Fig. 2. Application architecture of the treemap prototype.

When the users trigger the generation of a new set of recommendations or the alteration of the currently displayed treemap, the parameters of their request are transferred to the recommendation engine in the form of a JSON document. The parameters mainly include the assignment of the attributes to the treemap dimensions (cf. Sect. 3) as well as the recommender algorithm along with the corresponding settings. Regarding the recommender algorithms, the prototype relies on the Vogoo PHP library[1], whose user-based collaborative filtering, item-based collaborative filtering and SlopeOne implementations are used in slightly adjusted forms. The recommendation engine collects the data needed to process the request from the database management system (DBMS), generates the set of recommendations, and forwards them to the treemap graphical user interface (GUI) in the form of a JSON document. Finally, this JSON document is employed to create the treemap, which is realized using the JavaScript libraries D3.js and D3plus.

---

[1] https://sourceforge.net/projects/vogoo/.

5   Descriptive Evaluation

To demonstrate the proper functioning and goodness of our solution, we carry out a descriptive scenario-based evaluation. In particular, we present a realistic scenario with which we show how our prototype can enhance the exploration of recommendations in practice. To construct the scenario, the prototype has been extended with the functionality to add new users and ratings to the dataset. Due to space restrictions, however, these procedures are not discussed in detail. Instead, the reader is referred to the following URL created for demonstration purposes: http://rec-ifs.uni-r.de/treemap/ (Note that the debug mode allows to use the tool from the perspective of each of the MovieLens 100K users.)

5.1   Scenario Analysis

Suppose it is Friday evening and Alice wants to watch a movie together with her boyfriend Bob. In general, Alice loves romances and likes comedies and dramas. She dislikes western, sci-fi and crime movies. Considering the presence of Bob, Alice is quite sure that she wants to receive a recommendation for a comedy or a drama for tonight. A romance would be conceivable only if it was a highly popular and generally liked one. In addition, any movie with a very high predicted rating for Alice might be a good choice as long as it is not from one of her disliked genres. Applying a genre filter before having seen any recommendations may exclude these other good alternatives simply because Alice is not aware of them. Hence, Alice decides to apply no genre filter at all before not having explored the search space with the help of our treemap visualization.

After Alice has provided the recommender system with several movie ratings, the treemap is generated using the default algorithm settings (user-based collaborative filtering, 15 nearest neighbors, 25 displayed recommendations) as depicted in Fig. 3. Because of the clustering of the movies by genre and the clearly depicted ratios between them, Alice quickly gets an impression of the relevance of the different genres. As expected from her personal preferences, romances are comparatively important. However, comedies and mystery movies are even more important in the search space than romances. Looking at single movies, the two mystery movies "Thin Man, The" and "Lone Star" as well as the comedies "Much Ado About Nothing" and "Shall We Dance? 1996" are particularly conspicuous because of the large size of the corresponding rectangles. As a result of this first overview of the search space, Alice now knows that especially comedies, romances, and mystery movies come into question for tonight.

With this insight in mind, it makes sense for Alice to compare the recommendations in more detail by applying different filters and displaying different attributes through the treemap dimensions. Hereto, she selects only "Comedy", "Mystery", and "Romance" in the genre filter above the treemap visualization. She also reduces the number of recommendations to 10, which makes the size differences of the rectangles even more obvious. Using the opacity dimension to visualize the release years of the movies reveals that "Thin Man, The" is from 1934.

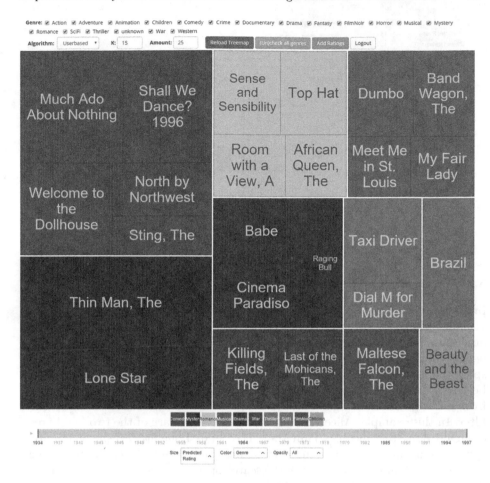

Fig. 3. Screenshot of the software prototype.

Although this movie has the highest predicted rating among the recommendations, Alice realizes that she prefers a more recent movie for tonight and decides to exclude older ones from her decision making. Hence, she sets the lower bound of the timeline filter below the treemap visualization to 1988. By visualizing the movies' overall ratings through the color dimension, Alice discovers that the comedy "Shall We Dance? 1996" has both a high overall and high predicted rating and would therefore be a suitable candidate. However, the tooltip information corresponding to the movie shows that its theme is very musical (cf. Fig. 4), which is something Bob does not like at all. The remaining alternatives with a very high predicted rating for Alice are the comedy "Much Ado About Nothing" and the mystery movie "Lone Star". As they do not show any notable differences regarding their attributes, Alice finally chooses "Much Ado About Nothing" because the tooltip information mentions that it is based on the identically named play by Shakespeare, which perfectly fits her preferences.

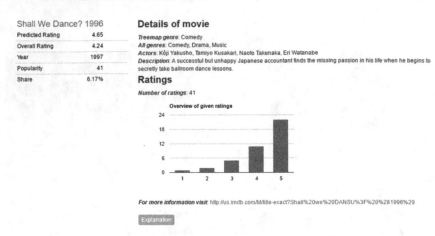

Fig. 4. Screenshot of the tooltip corresponding to "Shall We Dance? 1996".

5.2   Evaluation of Research Objectives

In summary, the scenario analysis demonstrates several major benefits of our visualization prototype, which are directly related to the accomplishment of the objectives pointed out in Sect. 2. First and most important, the users' comprehension of the recommendations is facilitated by educating them about the search space (cf. first objective). On the one hand, the structured initial overview of the best alternatives in the search space clustered by categories prevents the users from missing out on very good alternatives because of a rash usage of a category filter. In this example, Alice would have never been aware of the two highly recommended mystery movies "Thin Man, The" and "Lone Star" if she had limited her recommendations to romances, comedies, and dramas from the beginning. On the other hand, the overview of the search space – especially through the size dimension – helps the users get a quick impression of the amount of alternatives with a very high predicted rating. This insight can be used to reduce the number of simultaneously displayed recommendations, so that a more detailed analysis is limited to the ones that actually come into question.

Second, the users can interactively explore the recommendations with their situational needs in mind (cf. second objective). In addition to providing a structured overview, the treemap (along with its dimensions) enables the comparison of multiple attributes at once without reducing the visualization's readability. In the scenario, for example, we mention the simultaneous depiction of three attributes: the genre through hierarchy, the predicted rating through size, and the overall rating through color.

Finally, although not specific to our treemap visualization, the transparency of the recommender system is increased (cf. third objective) by providing the users with as many information available to the recommender system as possible, which includes both raw data from the dataset and intermediate values from the recommendation process. Thus, the users are supported in understanding the accomplishment of their recommendations.

6   Conclusion and Future Work

Even though today's recommender algorithms are highly sophisticated, they can hardly take into account the users' momentary situational needs. This is especially the case if these needs deviate notably from the users' normal preferences determined according to their past rating behaviors. An obvious way to address this is to initially inquire the users' situational needs. However, this may lead to the loss of several good alternatives because the users are unlikely to be able to accurately state all of their preferences upfront. Against this background, this paper presented a treemap visualization that does not require such additional input from the users and instead supports them in the explorative analysis of the recommended items with their situational needs in mind. In the descriptive scenario-based evaluation, we showed that the structured overview provided by the treemap could facilitate the users' comprehension of the big picture of alternatives available in the search space. This helps them understand the reasoning behind the recommendations and serves them as a basis to get clear about their situational needs, inspect the most relevant recommendations in detail, and finally arrive at informed decisions.

In future work, further insights might be gained by using the treemap visualization in application areas other than movies (e.g. books, music) and also with content-based algorithms. Moreover, it would be possible to extend our visualization tool with findings from related recommender systems research areas (e.g. active learning [39], critiquing [40], and more extensive explanations).

Acknowledgments. The research leading to these results was supported by the "Bavarian State Ministry of Education, Science and the Arts" as part of the FORSEC research association. The authors would like to thank Kilian Müller and Regina Staller for the implementation of the prototype used in this paper.

References

1. Hill, W., Stead, L., Rosenstein, M., Furnas, G.: Recommending and evaluating choices in a virtual community of use. In: Proceedings of the 13th SIGCHI Conference on Human Factors in Computing Systems (CHI), pp. 194–201 (1995)
2. Resnick, P., Iacovou, N., Suchak, M., Bergstrom, P., Riedl, J.: GroupLens: an open architecture for collaborative filtering of netnews. In: Proceedings of the 1994 ACM Conference on Computer Supported Cooperative Work (CSCW), pp. 175–186 (1994)
3. Resnick, P., Varian, H.R.: Recommender systems. Commun. ACM **40**(3), 56–58 (1997)
4. Adomavicius, G., Tuzhilin, A.: Toward the next generation of recommender systems: a survey of the state-of-the-art and possible extensions. IEEE Trans. Knowl. Data Eng. **17**(6), 734–749 (2005)
5. Konstan, J.A., Riedl, J.: Recommender systems: from algorithms to user experience. User Model. User Adapt. Interact. **22**(1–2), 101–123 (2012)
6. Herlocker, J.L., Konstan, J.A., Terveen, L.G., Riedl, J.T.: Evaluating collaborative filtering recommender systems. ACM Trans. Inf. Syst. **22**(1), 5–53 (2004)

7. Netflix: The Netflix prize rules. http://www.netflixprize.com/rules (2009). Accessed 19 Apr 2016

8. Loepp, B., Herrmanny, K., Ziegler, J.: Blended recommending: integrating interactive information filtering and algorithmic recommender techniques. In: Proceedings of the 33rd Annual ACM Conference on Human Factors in Computing Systems (CHI), pp. 975–984 (2015)

9. Pu, P., Chen, L., Hu, R.: Evaluating recommender systems from the user's perspective: survey of the state of the art. User Model. User Adapt. Interact. **22**(4–5), 317–355 (2012)

10. Knijnenburg, B.P., Willemsen, M.C., Gantner, Z., Soncu, H., Newell, C.: Explaining the user experience of recommender systems. User Model. User Adapt. Interact. **22**(4–5), 441–504 (2012)

11. McNee, S.M., Riedl, J., Konstan, J.A.: Being accurate is not enough: how accuracy metrics have hurt recommender systems. In: CHI 2006 Extended Abstracts on Human Factors in Computing Systems, pp. 1097–1101 (2006)

12. Swearingen, K., Sinha, R.: Beyond algorithms: an HCI perspective on recommender systems. In: Proceedings of the ACM SIGIR Workshop on Recommender Systems (2001)

13. Burke, R.: Knowledge-based recommender systems. In: Kent, A. (ed.) Encyclopedia of Library and Information Science, vol. 69 - Supplement 32, pp. 180–200. Marcel Dekker, Inc., New York (2000)

14. Burke, R., Hammond, K., Young, B.: Knowledge-based navigation of complex information spaces. In: Proceedings of the 13th National Conference on Artificial Intelligence (AAAI), pp. 462–468 (1996)

15. Herrmanny, K., Schering, S., Berger, R., Loepp, B., Günter, T., Hussein, T., Ziegler, J.: MyMovieMixer: Ein hybrider Recommender mit visuellem Bedienkonzept. In: Mensch und Computer 2014 - Tagungsband, pp. 45–54 (2014)

16. Tintarev, N., Masthoff, J.: A survey of explanations in recommender systems. In: Proceedings of the 23rd International Conference on Data Engineering Workshop (ICDE), pp. 801–810 (2007)

17. Smyth, B.: Case-based recommendation. In: Brusilovsky, P., Kobsa, A., Nejdl, W. (eds.) The Adaptive Web. LNCS, vol. 4321, pp. 342–376. Springer, Heidelberg (2007). doi:10.1007/978-3-540-72079-9_11

18. Hevner, A.R., March, S.T., Park, J., Ram, S.: Design science in information systems research. MIS Q. **28**(1), 75–105 (2004)

19. Peffers, K., Tuunanen, T., Rothenberger, M.A., Chatterjee, S.: A design science research methodology for information systems research. J. Manag. Inf. Syst. **24**(3), 45–77 (2007)

20. O'Donovan, J., Smyth, B., Gretarsson, B., Bostandjiev, S., Höllerer, T.: PeerChooser: visual interactive recommendation. In: Proceedings of the 26th SIGCHI Conference on Human Factors in Computing Systems (CHI), pp. 1085–1088 (2008)

21. Gretarsson, B., O'Donovan, J., Bostandjiev, S., Hall, C., Höllerer, T.: SmallWorlds: visualizing social recommendations. Comput. Graph. Forum **29**(3), 833–842 (2010)

22. Gou, L., You, F., Guo, J., Wu, L., Zhang, X.: SFViz: Interest-based friends exploration and recommendation in social networks. In: Proceedings of the 2011 International Symposium on Visual Information Communication (VINCI), pp. 1–10 (2011)

23. Bostandjiev, S., O'Donovan, J., Höllerer, T.: TasteWeights: a visual interactive hybrid recommender system. In: Proceedings of the 6th ACM Conference on Recommender Systems (RecSys), pp. 35–42 (2012)

24. Parra, D., Brusilovsky, P., Trattner, C.: See what you want to see: visual user-driven approach for hybrid recommendation. In: Proceedings of the 19th International Conference on Intelligent User Interfaces (IUI), pp. 235–240 (2014)
25. Gansner, E., Hu, Y., Kobourov, S., Volinsky, C.: Putting recommendations on the map: visualizing clusters and relations. In: Proceedings of the 3rd ACM Conference on Recommender Systems (RecSys), pp. 345–348 (2009)
26. Kagie, M., van Wezel, M., Groenen, P.J.: Map based visualization of product catalogs. In: Ricci, F., Rokach, L., Shapira, B., Kantor, P.B. (eds.) Recommender Systems Handbook, pp. 547–576. Springer, Boston (2011)
27. Verbert, K., Parra, D., Brusilovsky, P., Duval, E.: Visualizing recommendations to support exploration, transparency and controllability. In: Proceedings of the 18th International Conference on Intelligent User Interfaces (IUI), pp. 351–362 (2013)
28. Tintarev, N., Masthoff, J.: Designing and evaluating explanations for recommender systems. In: Ricci, F., Rokach, L., Shapira, B., Kantor, P.B. (eds.) Recommender Systems Handbook, pp. 479–510. Springer, Boston (2011)
29. Katarya, R., Jain, I., Hasija, H.: An interactive interface for instilling trust and providing diverse recommendations. In: Proceedings of the International Conference on Computer and Communication Technology (ICCCT), pp. 17–22 (2014)
30. Weskamp, M.: Newsmap. http://newsmap.jp/ (2010). Accessed 19 Apr 2016
31. Johnson, B., Shneiderman, B.: Tree-maps: a space-filling approach to the visualization of hierarchical information structures. In: Proceedings of the 2nd Conference on Visualization (VIS), pp. 284–291 (1991)
32. Shneiderman, B.: Tree visualization with tree-maps: 2-D space-filling approach. ACM Trans. Graph. 11(1), 92–99 (1992)
33. Gasteier, K., Krois, K., Hrdina, F.: Exploratory search and content discovery: the semantic media browser (SMB). In: Proceedings of the 13th International Conference on Knowledge Management and Knowledge Technologies (i-Know), pp. 1–8 (2013)
34. Gemignani, Z.: 10 Lessons in Treemap Design. http://www.juiceanalytics.com/writing/10-lessons-treemap-design (2009). Accessed 19 Apr 2016
35. Vig, J., Sen, S., Riedl, J.: Tagsplanations: explaining recommendations using tags. In: Proceedings of the 14th International Conference on Intelligent User Interfaces (IUI), pp. 47–56 (2009)
36. Sinha, R., Swearingen, K.: The role of transparency in recommender systems. In: Extended Abstracts on Human Factors in Computing Systems (CHI EA), pp. 830–831 (2002)
37. Herlocker, J.L., Konstan, J.A., Riedl, J.: Explaining collaborative filtering recommendations. In: Proceedings of the 2000 ACM Conference on Computer Supported Cooperative Work (CSCW), pp. 241–250 (2000)
38. Harper, F.M., Konstan, J.A.: The MovieLens datasets: history and context. ACM Trans. Interact. Intell. Syst. 5(4), 1–19 (2016)
39. Rubens, N., Elahi, M., Sugiyama, M., Kaplan, D.: Active learning in recommender systems. In: Ricci, F., Rokach, L., Shapira, B. (eds.) Recommender Systems Handbook, pp. 809–846. Springer, Boston (2015)
40. Chen, L., Pu, P.: Critiquing-based recommenders: survey and emerging trends. User Model. User Adapt. Interact. 22(1–2), 125–150 (2012)

Data Management and Data Analysis

An E-Shop Analysis with a Focus on Product Data Extraction

Andrea Horch$^{(\boxtimes)}$, Andreas Wohlfrom, and Anette Weisbecker

Fraunhofer Institute for Industrial Engineering IAO, 70569 Stuttgart, Germany
{andrea.horch,andreas.wohlfrom,anette.weisbecker}@iao.fraunhofer.de
https://www.iao.fraunhofer.de

Abstract. E-commerce is a constantly growing and competitive market. Online prices are updated daily or even more frequently, and it is very important for e-shoppers to find the lowest price online. Therefore, e-shop owners need to know the prices of their competitors and must be able to adjust their own prices in order to remain competitive. The manual monitoring of all prices of all products and competitors is too time-consuming; hence, the e-shop owners need software support for that task. For the development of such software tools the developers need a profound comprehension of the structure and design of e-shop websites. Existing software tools for Web data extraction are based on the findings of different website analyzes. The existing tools show: The more specific and detailed the analysis and the analyzed websites, the better the data extraction results. This paper presents the results and the derived findings of a deep analysis of 50 different e-shop websites in order to provide new insights for the development and improvement of software tools for product data extraction from e-shop websites.

Keywords: E-shop statistics · Product data extraction

1 Introduction

E-commerce is a huge and constantly growing market. In 2015 the turnover of the European e-commerce increased by 14.3 % to 423.8 billion Euros. There were 331 million e-shoppers and an estimated amount of more than 715,000 online businesses [1]. These numbers suggest that there is a high competition between European e-shops. According to [2], online prices are daily or even more frequently updated and [3] show that it is very important for e-shoppers to find the lowest price online knowing the prices of the competitors, and being able to adjust their own prices is essential for e-shop owners to remain competitive.

To be able to monitor the prices of a large amount of products which are only available within the websites of the competitors, e-shop owners need to be supported by software tools. Such software tools need to implement approaches for (1) the automated detection of product offers within the e-shop websites, (2) the automated identification and extraction of defined product attributes (e.g. product name or product price) within the product offers, as well as (3) the matching

© Springer International Publishing AG 2017
D. Bridge and H. Stuckenschmidt (Eds.): EC-Web 2016, LNBIP 278, pp. 61–72, 2017.
DOI: 10.1007/978-3-319-53676-7_5

of identical and similar products of two or more e-shops in order to compare the products and their prices. For the implementation of these approaches a profound comprehension of the structure and design of e-shop websites and the contents is needed.

This paper provides the necessary knowledge through presenting and examining the results of a deep analysis of the structure and contents of 50 different e-shop websites as well as by drawing conclusions from these results in order to exploit the findings for the extraction of product data from e-shop websites.

2 Related Work and Motivation

A popular approach for the extraction of data records from Web pages is called MDR (Mining Data Records in Web pages). The MDR algorithm is presented in [4]. The approach is based on two findings about Web pages containing data records: (1) Groups of data records describing akin objects use similar HTML tags and are located within contiguous regions of a Web page, and (2) these groups usually have the same parent node within the HTML tree of the Web page.

ClustVX introduced in [5,6] exploits the structural and visual similarity of template-generated Web pages for the extraction of data records. Similar to MDR ClustVX is based on the observation that the essential data records of a Web page can be identified by a common parent node, and therefore these data records have almost the same XPath expression[1]. ClustVX complements this basis with the finding that template-generated data records additionally include repeating visual patterns which are stored in the CSS (Cascading Style Sheets) information of the website.

Machine learning approaches for product feature extraction as the unsupervised information extraction system OPINE presented in [7] are usually based on natural language processing (NLP) and user-generated example data. Therefore, such methods depend on a specific product domain and a particular language. Additionally, the generation of example data is very time-consuming. The techniques used by approaches as MDR or ClustVX are more flexible as they are based on the structure of the website instead of the textual content, and thus they are independent of a product domain or language. MDR and ClustVX are successful approaches for Web data record extraction which are based on the findings derived from the results of Website analyses.

In [8–13] we present our previous work on automated product record and product attribute extraction from e-shop websites. Our work is based on previous website analyses and achieves good results. As our approach analyzes the website structure to identify product records, and the algorithms use structural knowledge and patterns to find the product attributes, the approach is independent from a specific product domain or a particular website language. To improve further our algorithms we started another analysis of e-shop websites which is

---

[1] Compound tag path from the root node of the website's HTML tree to a particular element of the Web page.

presented in this paper. The results of the analysis shall support developers of approaches for product data record extraction by providing useful information and hints for the automated identification and extraction of product records and product attributes from e-shop websites.

3   Analyzed E-Shops

We have manually analyzed an overall of 50 e-shops from five different countries: ten e-shops each from Germany, Spain, United Kingdom, Greece and the United States. The distribution of product categories of the analyzed e-shops is shown in Fig. 1. The e-shops were randomly selected from the results of different keyword searches consisting of the keyword "shop" followed by the product category and the country. The searches were performed by using the Google search engine.

The product categories are a combination of the most popular product categories as stated in [2] and rather uncommon categories picked out of the set of pilot e-shops of our last research project. In Fig. 1 the most popular product categories presented in [2] are marked by an asterisk (*). The product categories of the e-shops are distributed according to the amount of subcategories as, for example, the e-shops of the category *clothing* include e-shops specialized on clothes or footwear, and those of the category *cosmetics &haircare* contain shops selling haircare, skincare or cosmetics. The product categories which are represented by only one e-shop are very special categories. The set of e-shops for analysis is a mixture of small, medium and large e-shops. Small e-shops are shops offering less than 1,000 different products, medium-sized e-shops sell between 1,000 and 10,000 products, and large e-shops have a product range of

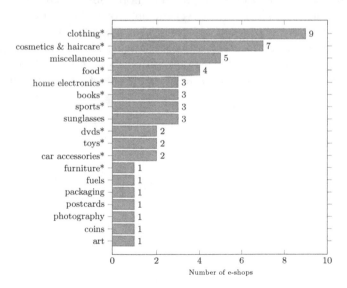

Fig. 1. Category distribution of the e-shops for analysis.

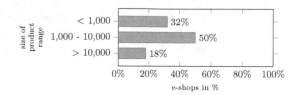

Fig. 2. Distribution of e-shops based on the size of their product range.

more than 10,000 different products. The distribution of e-shops based on the size of offered products (product range) is presented in Fig. 2.

The distribution of the analyzed e-shops based on the size of their product ranges was composed randomly. But with a distribution of 32% of small e-shops, 50% of medium-sized e-shops, and 18% of large e-shops we expect to have a representative distribution for e-shops that are not too big, and hence the big e-shops are not the main focus of this work.

4   Pages Including Product Lists

In order to find and extract the lists of product offers within an e-shop website the necessary crawling depth need to be determined to avoid crawling the whole e-shop website which can consume a lot of time and computing resources. For the determination of the required crawling depth we have analyzed the 50 e-shop websites according to the location of the lists of product offers. The results of this analysis are presented in Fig. 3.

Figure 3 shows that 78% of the analyzed e-shop websites have lists of product offers directly on their start page (level 0). 96% of the e-shops include list of product offers within the pages of website level 1, which are Web pages available through clicking on a link located on the start page. 78% of the e-shops contain lists of product offers in level 2, which are all pages accessible through links on the Web pages of level 1 (one click away from the start page). Only 12% of

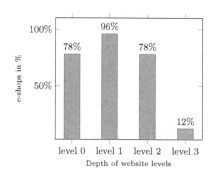

Fig. 3. Distribution of pages including product lists within e-shop website.

the analyzed e-shop websites have product offer lists on level 3, which are Web pages available through a link on level 2 (two clicks away from the start page). We found out that the product lists offered on level 0 (start page) are special offers of the e-shop which appear more frequently on the different levels of the e-shop website. The product lists available on level 1 are the full list of products or the first page of the full product offer list. The product lists shown on level 2 are further pages of the product list beginning on level 1. E-shops including product offer list on level 3 integrate a second level of product categories on level 1, that means the actual product offer list begins on level 2 instead of level 1 and is continued on the pages of level 3. There was no e-shop website within the set of analyzed e-shops which presented product offer list on a deeper level of the website than level 3. Therefore, all products of the analyzed e-shops could be found by crawling the website until level 3. Another finding of our analysis is that independent from specific search buttons or search fields offered on an e-shop website, the product lists of the website can be simply found by following all links of the website domain on the e-shop website until those on level 3.

5   Product Offers

To gain knowledge about the structure of product offers as well as the location of those within the e-shop websites we have analyzed 25 products per e-shop website from two to five different product list pages of the respective e-shop. The number of pages the product offers were taken from was depended on the amount of products on a product list page of the e-shop.

The first analysis concerning the product offers provides information about the most common HTML tags of the e-shop Web pages the product offers are embedded in. Figure 4 presents the distribution of HTML tags embedding the product offers of the lists of the analyzed e-shops. 52% of the analyzed product offers are embedded into *div* tags, 36% of the e-shops use tags for list items (*li*) to integrate their product offers within their Web pages. The other 12% use other tags as table cells (*td*), whole HTML tables (*table*) or anchor tags (*a*) to embed their product offers into the HTML trees of their Web pages. The analysis shows that it is very likely (88%) that the product offers of an e-shop website are

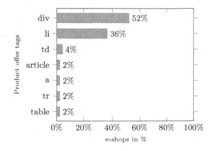

Fig. 4. Distribution of product offer tags.

included into *div* or *li* tags, but there is also a chance that they are embedded into another kind of tag. For a 100% coverage of e-shop websites developers of web data extraction algorithms need to consider all possible kinds of tags in order to be able to find the product offers.

Another analysis concerns the most frequent number of parent HTML elements of the product elements of the e-shops. Namely, within how many elements of the HTML tree of the e-shop website the product offer elements are embedded. Figure 5 shows the results of this analysis for the 50 analyzed e-shops.

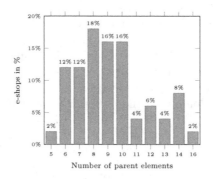

Fig. 5. Most frequent number of parents of product offers.

The product offer elements of the analyzed e-shops have a range of parent elements from 5 to 16 parents. Most product offers (74%) are embedded into 6 to 10 parent elements. We could observe that within one Web page the count of parent elements for all product offers is the same, but for different Web pages of an e-shop the amount of parent elements of the product offers of the different pages can vary. This is caused by some special regions as advertising elements which are only included into some pages of the e-shop website. The main finding of this analysis is that all product offers of a Web page have the same parent elements, and therefore they have the same tag path from the root element of the HTML page tree (*html* tag) to the product element itself.

The most frequent number of child elements of the product offer elements was also determined for the analyzed e-shops. Figure 6 shows the results.

The most frequent number of child elements of the product offers within the e-shops has a range from 3 to 70 child elements. This number is very distributed and therefore it is impossible to derive anything from these results. Additionally, the number of child elements of the different e-shops varies widely within one Web page of an e-shop as some product offers of the page include additional elements such as an offer price or a product rating which others do not.

Further, we have analyzed the most frequent number of image elements contained by the product offers. The results are presented in Fig. 7.

The most product offers (70%) integrate exactly one product image. One of the e-shops did not contain any image in the majority of product offers. The

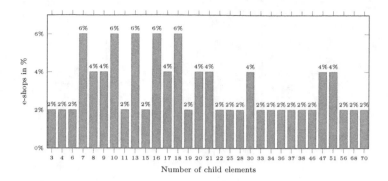

Fig. 6. Most frequent number of children of product offers.

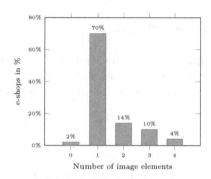

Fig. 7. Most frequent number of image elements within product offers.

others include 2 to 4 images as the product image, detail view images of the product for the hover effect of the product image, and images showing the star rating. There are e-shops which integrate up to 11 images in their product offers. The insight of this analysis is that usually one image or more are included within the product offers of e-shops, but not all images within the product offers show the product, and there is a small number of product offers which do not contain any image.

Next, we have determined the most frequent number of anchor elements that product offers contain (see Fig. 8).

Most product offers (66%) include 2 to 4 anchor elements. Most of these anchors link to the detail page of the corresponding product, but there are also links connecting to further information as shipping prices or user ratings. There is also a case when one of the e-shops does not contain any anchor within its product offers as the product offer element itself is an anchor element. We could even identify product offers which contain more than 20 anchors. The findings of this analysis are that the product offers of e-shops contain at least one anchor element (linking to the detail page of the product) or the product offer element itself is an anchor (also linking to the detail page).

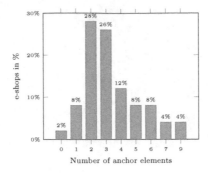

Fig. 8. Most frequent number of anchor elements within product offers.

6   Attributes of Product Offers

To provide data required for the identification and extraction of the attributes of the product offers as product name, product prices or product image we have collected structural and visible information about the product attributes of the 50 analyzed e-shops.

6.1   Product Image

98% of the analyzed e-shops use an *img* tag to integrate the product image, only 2% (1 e-shop) utilize anchor tags. 96% (48 e-shops) show images in JPEG (Joint Photographic Experts Group) format whereof 47 e-shops use the file extension "jpg" and 1 e-shop puts the "jpeg" extension. The other 4% use PNG (Portable Network Graphics) images. In 100% of the cases the product image was the biggest image in the product offer element, and in 98% the product image was the first image within a product offer element. Hence, for the identification and extraction of the product image of a product offer the first and the biggest image(s) should be considered. For the localization of images in product offers the image (*img*) and anchor (*a*) tags as well as the search for the file extensions *jpeg*, *jpg* and *png* can be useful.

6.2   Product Name

We have analyzed in what kind of tag elements the product name can be found within product offers. In 70% of the analyzed e-shops the product name was embedded into an anchor tag (*a*). In other 10% the name of the product could be found within a *span* element. In the rest of the cases the product name was found in heading tags (*h2* to *h4*), division tags (*div*), *strong* tags or paragraph tags (*p*). In 98% of the e-shops the product name is a text within an HTML tag. In 24% of the cases the product name can be found as a *title* attribute. The product name is the text with the biggest font size (CSS information) within an anchor tag in 80% of the analyzed e-shops. In 76% the product name is the *alt* attribute of the product image element, in 20% it is the *title* attribute of the product image element, and in 82% it is the *alt* or *title* attribute of the product image element. Following this findings it is very likely to find the product name more than once within the product offers, and most likely it is the *title* or *alt* attribute of the product image element.

6.3   Product Prices

There can be found many different prices within one product offer. In order to identify the right values for the actual product price or the regular price of the product we have analyzed the tags, the structure, and the visual attributes of the prices of the product offers within the 50 analyzed e-shops. Table 1 shows different 15 types of prices we could identify within the product offers during our analysis.

Table 1. Price Types of E-Shop Offers

No.	Price type	No.	Price type
1	Product price (incl. tax)	9	Tax
2	Product price (excl. tax)	10	Price for shipping
3	Regular price	11	Min. price of purchase for free shipping
4	Price of voucher for purchase	12	Deposit for product container
5	Price per unit (e.g. 100 g)	13	Start price for bids
6	Price range	14	Quantity price
7	Secret product price	15	Open box price
8	Savings		

In many e-shops there is information about the product prices including and excluding tax or the tax is stated separately (see no. 1, 2 and 9). Regular product price (see no. 3) means the price before a certain price reduction of the product offer. There are prices of vouchers (see no. 4) which can be purchased to get a price reduction. For some products a unit price (see no. 5) as price per gram is given. Some prices have a product range (see no. 6) as the price depends on an ordered quantity or a size. Some e-shops provide a secret product price (see no. 7) as their price for a product lies under the price allowed by the manufacturer. For special offers the savings amount (see no. 8) is presented. In many cases the price for shipping (see no. 10) the product is declared within the product offer, and some e-shops offer free shipping for orders from a minimum amount (see no. 11). Some products can only be sold within a container (e.g. gases) and the customer needs to pay a deposit for that container (see no. 12). On platforms like eBay there is a start price for bids instead or besides a product price (see no. 13). Some products are sold by quantity, and therefore a quantity price (see

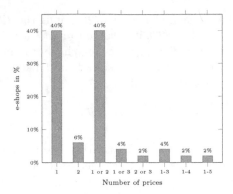

Fig. 9. Number of prices within the analyzed e-shops.

no. 14) is declared. Another type of a price called open box price is a special offer price for used products or display items (see no. 15).

As shown in Fig. 9 the product offers of the analyzed e-shops contain between 1 and 5 different prices of the price types given in Table 1. Most product offers include 1 or 2 prices (80%).

We have also analyzed the distribution of HTML elements including the prices of online price offers. Most e-shops (84%) embed their prices into *span*, division (*div*) or anchor tags (*a*), but there are many other possible tags a price can be embedded in, thus the price of a product offer cannot be found by searching for a specific kind of tag.

The most common visual differences between the actual and the regular price of an online price offer defined by Cascading Style Sheets (CSS) are shown in Fig. 10. These visual differences are: (1) the font color, (2) the text decoration as regular prices most often are crossed-out (CSS value *line-through* or HTML

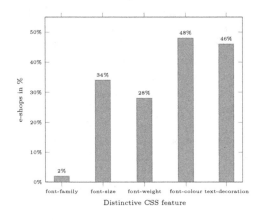

Fig. 10. Differences of actual and regular product price.

elements *del* or *strike*), (3) the font size or (4) the font weight. In the graph of Fig. 10 the e-shops that do not contain a regular price are also counted. But in 50% of the e-shops which include a regular price (15 of 30 e-shops) the font-size of the actual price was bigger than the font-size of the regular prices. Therefore, the CSS of the page should be considered and exploited in order to identify the right prices within the product offers.

7   Conclusions

We have analyzed 50 e-shop websites selling different product categories and 25 product offer records of each e-shop (1,250 records) for their structural and visual composition for the purpose of Web data extraction. The findings of the analysis provides the required knowledge about the design of e-shop websites to the developers of services for the automated extraction of product offers from e-shop websites. The main findings of our analysis are the following:

1. The necessary crawling depth to find all product lists of an e-shop website is level 3, which are pages three clicks away from the start page of the website.
2. Most e-shops include at least one image in their product offers. Usually, the biggest image is the product image.
3. There may be different prices within a product offer. We could identify 15 different kinds of prices. In most cases the regular price was crossed out by a "line-through" text-decoration property in the CSS of the page, and the actual product price of the product was visually different from other prices in font size, font color or font weight. Thus, it is advisable to include the CSS of the page into the analysis in order to be able to distinguish the prices.

References

1. Ecommerce Foundation: European B2C E-commerce Report 2015 (Light Version). Report, Ecommerce Foundation (2015)
2. Civic Consulting: Consumer market study on the functioning of e-commerce and Internet marketing and selling techniques in the retail of goods. Report, Civic Consulting (2011)
3. PostNord: E-commerce in Europpe 2015. Report, PostNord (2015)
4. Liu, B., Grossman, R., Zhai, Y.: Mining data records in web pages. In: Proceedings of the Ninth ACM SIGKDD International Conference on Knowledge Discovery and Data Mining (KDD-2003), pp. 601–606. ACM, New York (2003)
5. Grigalis, T.: Towards automatic structured web data extraction system. In: Local Proceedings and Materials of Doctoral Consortium of the Tenth International Baltic Conference on Databases and Information Systems, pp. 197–201, Vilnius (2012). CEUR-WS.org
6. Grigalis, T., Čenys, A.: Unsupervised structured data extraction from template-generated web pages. J. Univ. Comput. Sci. (J.UCS) **20**(2), 169–192 (2011)
7. Popescu, A.-M., Etzioni, O.: Extracting product features and opinions from reviews. In: Proceedings of the Conference on Human Language Technology and Empirical Methods in Natural Language Processing, pp. 339–346. Association for Computational Linguistics, Stroudsburg, PA, USA (2005)

8. Horch, A., Kett, H., Weisbecker, A.: A lightweight approach for extracting product records from the web. In: Proceedings of the 11th International Conference on Web Information Systems and Technologies, pp. 420–430. SciTePress (2015)

9. Horch, A., Kett, H., Weisbecker, A.: Mining E-commerce data from E-shop websites. In: 2015 IEEE TrustCom/BigDataSE/ISPA, vol. 2, pp. 153–160. IEEE (2015)

10. Horch, A., Kett, H., Weisbecker, A.: Extracting product unit attributes from product offers by using an ontology. In: Second International Conference on Computer Science, Computer Engineering, and Social Media, CSCESM 2015, pp. 67–71. IEEE (2015)

11. Horch, A., Kett, H., Weisbecker, A.: Extracting product offers from e-shop websites. In: Monfort, V., Krempels, K.-H., Majchrzak, T.A., Turk, Ž. (eds.) WEBIST 2015. LNBIP, vol. 246, pp. 232–251. Springer, Heidelberg (2016). doi:10.1007/978-3-319-30996-5_12

12. Horch, A., Kett, H., Weisbecker, A.: Matching product offers of e-shops. In: Cao, H., Li, J., Wang, R. (eds.) PAKDD 2016. LNCS (LNAI), vol. 9794, pp. 248–259. Springer, Heidelberg (2016). doi:10.1007/978-3-319-42996-0_21

13. Horch, A., Kett, H., Weisbecker, A.: A web application for the collection and comparison of E-shop product data. In: 2016 6th International Workshop on Computer Science and Engineering, WCSE 2016, pp. 167–179. International Workshop on Computer Science and Engineering (WCSE) (2016)

The WDC Gold Standards for Product Feature Extraction and Product Matching

Petar Petrovski[✉], Anna Primpeli, Robert Meusel, and Christian Bizer

Data and Web Science Group, University of Mannheim, Mannheim, Germany
petar@informatik.uni-mannheim.de

Abstract. Finding out which e-shops offer a specific product is a central challenge for building integrated product catalogs and comparison shopping portals. Determining whether two offers refer to the same product involves extracting a set of features (product attributes) from the web pages containing the offers and comparing these features using a matching function. The existing gold standards for product matching have two shortcomings: (i) they only contain offers from a small number of e-shops and thus do not properly cover the heterogeneity that is found on the Web. (ii) they only provide a small number of generic product attributes and therefore cannot be used to evaluate whether detailed product attributes have been correctly extracted from textual product descriptions. To overcome these shortcomings, we have created two public gold standards: The WDC Product Feature Extraction Gold Standard consists of over 500 product web pages originating from 32 different websites on which we have annotated all product attributes (338 distinct attributes) which appear in product titles, product descriptions, as well as tables and lists. The WDC Product Matching Gold Standard consists of over 75 000 correspondences between 150 products (mobile phones, TVs, and headphones) in a central catalog and offers for these products on the 32 web sites. To verify that the gold standards are challenging enough, we ran several baseline feature extraction and matching methods, resulting in F-score values in the range 0.39 to 0.67. In addition to the gold standards, we also provide a corpus consisting of 13 million product pages from the same websites which might be useful as background knowledge for training feature extraction and matching methods.

Keywords: e-commerce · Product feature extraction · Product matching

1 Introduction

The Web has made it easier for organizations to reach out to their customers, eliminating barriers of geographical location, and leading to a steady growth of e-commerce sales.[1] Beside of e-shops run by individual vendors, comparison

---

[1] Retail e-commerce sales worldwide from 2014 to 2019 - http://www.statista.com/statistics/379046/worldwide-retail-e-commerce-sales/.

© Springer International Publishing AG 2017
D. Bridge and H. Stuckenschmidt (Eds.): EC-Web 2016, LNBIP 278, pp. 73–86, 2017.
DOI: 10.1007/978-3-319-53676-7_6

shopping portals which aggregate offers from multiple vendors play a central role in e-commerce. The central challenge for building comparison shopping portals is to determine with high precision which e-shops offer a specific product. Determining whether two offers refer to the same product involves extracting a set of features (product attributes) from the web pages containing the offers and comparing these features using a matching function. The extraction of detailed product features from the HTML pages is challenging, as a single feature may appear in various surface forms in headlines, the product name, and free-text product descriptions. Product matching is difficult as most e-shops do not publish product identifiers, such as *global trade item number* (GTIN) or ISBN numbers, but heterogeneous product descriptions have different levels of detail [8].

To evaluate and compare product matching methods a comprehensive gold standard is needed. The most widely known public gold standard for product matching was introduced by Köpcke et al. [3]. However, this gold standard has two shortcomings: First, the gold standard only contains offers from four sources (Amazon.com, GoogleProducts, Abt.com and Buy.com) and thus only partly covers the heterogeneity of product descriptions on the Web. Moreover the gold standard contains only four attributes: product title, description, manufacturer and price; with more detailed product attributes (such as screen size or amount of memory) being part of free-text product titles and descriptions. These attributes need to be extracted from the free-text before they can be exploited by sophisticated matching methods. A more recent gold standard for product matching was introduced by Ristoski and Mika [14]. Their gold standard contains offers from a large number of websites which employ Microdata and schema.org markup. Their gold standard thus overcomes the shortcoming that data is gathered only from a small number of e-shops. However, their gold standard provides only two textual product attributes (name and description) and can thus not be used to evaluate feature extraction methods.

In order to overcome both above mentioned shortcomings, this paper presents two publicly accessible gold standard datasets and a product data corpus which can be used to train and evaluate product feature extraction and product matching methods:

Gold Standard for Product Feature Extraction containing   over   500 annotated product web pages. On each web page, we manually annotated all product features which appear within: (i) the name of the product marked up with Microdata, (ii) description of the product marked up with Microdata, (iii) specification tables, and (iv) specification lists.

Gold Standard for Product Matching containing over 75 000 correspondences (1 500 positive, and 73 500 negative) between products from a product catalog, containing 150 different products from three different product categories, and products described on web pages.

Product Data Corpus containing over 13 million product-related web pages retrieved from the same web sites. This corpus might be useful as background knowledge for the semi-supervised training of feature extraction and matching methods.

All artefacts presented in this paper as well as the detailed results of the experiments are published as part of the WebDataCommons (WDC) project[2] and can be downloaded from the WDC product data corpus page[3].

The paper is structured as follows: Sect. 2 describes the selection of the websites and the products that are used for the gold standards. In Sects. 3 and 4 the creation of the two gold standard datasets is described and statistics about the datasets are presented. Section 5 briefly describes the corpus consisting of product web pages and the way it was crawled from the Web. The baseline approaches and their results based on the corresponding gold standard are presented in the subsequent section. The last section gives an overview of related work.

2 Product Selection

In the following, we describe the process which was applied to select the products used in the gold standards. Namely, we explain how the products from the three different product categories: *headphones, mobile phones* and *TVs.* were selected.

Table 3 shows the 32 most frequently visited shopping web sites, based on the ranking provided by *Alexa*[4], which we use for the product selection. We collected first the ten most popular products from the different web sites, for each of the three chosen product categories. We further complemented this list by similar products (based on their name). As example, we found the product *Apple iPhone 6 64 GB* to be one of the most popular amongst all shopping web sites. We therefore included also the products *Apple iPhone 6 128 GB* as well as *Apple iPhone 5* into our product catalog. Especially for the product matching task, this methodology introduces a certain level of complexity, as the product names only differ by one or two characters. All in all, for each product category we selected 50 different products.

3 Gold Standard for Product Feature Extraction

This section describes the process that we used to create the gold standard for product feature extraction from product web pages. First, we explain how the gold standard was curated and then state statistical insights.

3.1 Gold Standard Curation

We randomly selected 576 web pages, each containing a product description for one of the products selected in Sect. 2, from the product corpus detailed in Sect. 5. From the 576 product descriptions, 158 are belonging to the headphones category, 254 to the phones category and 164 to the TVs category.

---

[2] http://webdatacommons.org.
[3] http://webdatacommons.org/productcorpus/.
[4] http://www.alexa.com/.

From each page we identified four key sources of attributes: As we have already shown in former research [12], information annotated using the markup language Microdata[5] has proven to be a good source of product features. Making use of the open-source library *Any23*[6], we extracted the product `name`, as well as the product `description`, marked up with Microdata with the schema.org properties `schema:name` and `schema:description` from each product web page. Further, as the research presented in [13] has shown promising results extracting features from tables and lists, we used a similar approach to identify specification lists and specification tables on the product pages.

For each extracted source, we label the contained features with an appropriate feature name. As example, if the name of the product is the string *Apple iPhone 6*, we label the sub-string *Apple* as `brand` and *iPhone 6* as `model`. Two independent annotators in parallel annotated the web pages. In case of a conflict, a third annotator solved them.

We also mapped the list of annotated product features to the list of features contained in our product catalog (see Sect. 4.1). This mapping as well as the gold standard dataset is available on the gold standard web page.

3.2   Distribution of Annotated Features

In total, we were able to annotate 338 distinct features. Table 1 presents the frequency of properties per category for each of the labeled sources of attributes: Microdata name, Microdata description, specification table and specification list. The percent of frequency distribution is calculated from the total number of products of a product category. The table does not include a comprehensive list of the properties, but selects only those commonly occurring in each of the different annotation tasks. For the title and description we found a lot of `tagline` properties. Tagline was used for properties, which are not product specification related. As an example, when we found the title *amazing iPhone*, the sub-string *amazing* is annotated with the property `tagline`. Moreover, expected properties like `model`, `brand` and `product_type` can be seen amongst the top. For the specification table and specification list a relatively low frequency of properties, with even distribution, can be seen in the three different categories, suggesting a diversity of descriptors used by vendors.

The findings underline that features extracted from the four sources of product web pages contain valuable feature information. The identification of those features with a high precision is essential in order to perform further integration tasks, like the matching of products.

4   Gold Standard for Product Matching

In this section, we describe the process which was applied to curate the gold standard for product matching. Further we present valuable statistics about the created gold standard.

---

[5] https://www.w3.org/TR/microdata/.
[6] http://any23.apache.org/.

Table 1. Feature distribution on the labeled web pages, grouped by product category and labeled source

Headphones

Microdata name		Microdata description		spec. table		spec. list	
prop.	% Freq	prop.	% Freq	prop.	% Freq	prop.	% Freq
product_type	94.30	tagline	36.23	brand	89.51	impedance	91.97
brand	84.18	model	34.78	condition	89.51	frequency_response	91.08
model	81.01	product_type	33.33	mpn	89.51	sensitivity	81.26
tagline	65.82	brand	33.33	product_gtin	73.72	cable_length	47.33
condition	51.27	headphones.form_factor	28.99	model	47.39	package_weight	41.07
color	39.24	color	24.64	additional_features	42.12	headphones_technology	29.47
headphone_use	25.32	headphones_technology	24.64	color	42.12	weight	29.47
headphones_form_factor	24.68	additional_features	20.29	impedance	36.86	color	22.32
compatible_headphone_type	18.99	product_gtin	17.39	headphones.cup_type	36.86	connectivity_technology	19.64
compatible_headphone_brand	18.35	jack_plug	17.39	headphones.form_factor	31.59	max_input_power	17.86

Mobile Phones

title		description		spec. table		spec. list	
prop.	% Freq	prop.	% Freq	prop.	% Freq	prop.	% Freq
brand	87.80	brand	53.35	memory	52.47	brand	87.30
phone_type	81.50	phone_type	45.34	brand	49.19	product_type	86.20
tagline	78.74	tagline	45.34	phone_type	37.71	modelnum	85.09
computer_operating_system	73.62	computer_operating_system	44.01	color	26.24	compatible_phones	79.57
display_size	71.26	color	42.68	display_size	24.60	function	71.28
product_type	63.78	product_type	42.68	mpn	22.96	retail_package	70.73
rear_cam_resolution	51.57	compatible_phones	40.01	rear_cam_resolution	21.32	material	70.18
compatible_phones	36.61	rear_cam_resolution	30.67	phone_carrier	19.68	memory	69.62
network_technology	19.29	network_technology	30.67	condition	19.68	package_weight	67.97
display_size	18.11	display_size	29.34	computer_operating_system	18.04	package_size	64.65

TVs

title		description		spec. table		spec. list	
prop.	% Freq	prop.	% Freq	prop.	% Freq	prop.	% Freq
product_type	83.77	display_type	24.14	brand	76.33	viewable_size	74.83
tagline	79.87	brand	20.69	display_type	73.70	hdmi_ports	60.41
brand	79.22	tagline	17.24	mpn	55.28	3d_technology	59.51
display_type	70.13	model	17.24	viewable_size	44.75	usb	52.29
model	66.23	video_signal_standard	17.24	condition	42.11	image_aspect_ratio	49.59
viewable_size	48.70	product_type	17.24	display_resolution	42.11	computer_operating_system	36.07
total_size	45.45	smart_capable	17.24	model	34.22	depth	34.26
refresh_rate	20.13	display_resolution	17.24	product_type	23.69	hdmi	29.75
display_resolution	19.48	total_size	17.24	weight	13.16	height	29.75
compatible_tv_type	18.83	curved	17.24	color	10.53	height_with_stand	29.75

In order to generate the gold standard for product matching, we create a product catalog containing the same 150 products described in Sect. 2. Moreover we make use of the web pages, we crawled based on the names of the products (Sect. 3).

4.1   Product Catalog

To complement the products with features, for each of the 150 products in our product catalog we obtained product-specific features from the manufacturers' web site or from *Google Shopping*.

Figure 1 shows two example pages, which we used to manually extract the features for our product catalog. Figure 1a depicts a product page on *Google Shopping*. While Fig. 1b depicts the manufacturers' web site for the same product.

(a) *Features provided by Google Shopping*   (b) *Features provided by the Manufacturer*

Fig. 1. Example of web pages from which we extracted data for the catalog

4.1.1   Availability of Product Features

In total, 149 different features are identified. We found 38 for products of the category *headphones*, 33 for the category *mobile phones*, and 78 for the category *TVs*.

Table 2 shows the availability of the number identified features for the products for each of the three categories, as well as showing some examples for each identified group. We find that, especially for TVs, 40 of the features are available for at least 50 % of the products. For the other products of the other two product categories, we found roughly around 20 features to be available for at least 50 % of the products. A description with the complete distribution of properties can be found on our web page.

Table 2. Density of the product features for the products contained in the product catalog and example features

	Number of features					
	10 | 20	30	40		50 60	70
Headphones	> 50% filled	20–50% filled	< 20% filled		N \A	
	form_factor	color	detach_cable			
	freq_response	magnet_mat	foldable			
	product_name	microphone	max_in_power			
	product_type	cup_type	height			
	conn._technology	diaphragm	width			
Mobile Phones	> 50% filled		20–50% filled		N\A	
	processor_type		core_count			
	display_resolution		product_code			
	display_size		manufacturer			
	height		package_height			
	product_name		modelnum			
TVs	>50% filled				20–50% filled	< 20% filled
	product_name				dlna	memory
	total_size				timer_functions	consumption
	hdmi_ports				screen_modes	response_time
	speakers_qty				pc_interface	brightness
	display_resolution				3d	batteries_inculded

4.2   Gold Standard Curation

We manually generated 1 500 positive correspondences, 500 for each product category. For each product of the product catalog at least one positive correspondence is included. Additionally, to make the matching task more realistic the annotators also annotate closely related products to the once in the product catalog like: phone cases, TV wall mounts or headphone cables, ear-buds, etc. Furthermore we created additional negative correspondences exploiting transitive closure. As all products in the product catalog are distinct, we can generate for all product descriptions contained in the web pages, where a positive correspondence exist to a product in the catalog, for all other products in the catalog a negative correspondence to this product on the web page.

Using the two approaches we ended up with 73 500 negative correspondences.

4.2.1   Distribution of Correspondences

The gold standard for product matching contains 75 000 correspondences, where 1 500 are correct.

Figure 2 depicts the number of positive correspondences which are contained for each product from the three different product categories. Evidently, more than 50% of the products have two or less correspondences. While only a few of the products have between 20 and 25 correspondences.

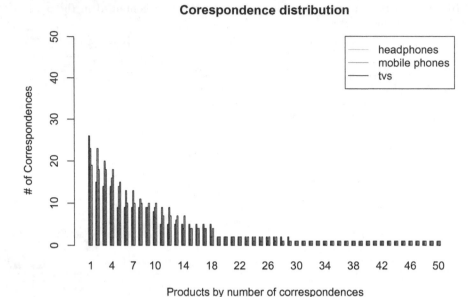

Fig. 2. Distribution of positive correspondences per category

5   Product Data Corpus

In addition to the two gold standard datasets, we have crawled several million web pages from 32 selected shopping web sites. Although we did not label all these web pages, we provide them for download as background knowledge for the semi-supervised training of feature extraction, product matching. or product categorization methods (Meusel et al. [9]).

Besides the keyword/product-based crawls which were already mentioned in Sect. 2, we performed a directed crawl for the 32 selected web sites, without the restriction to search for the specific products. We used the python-based, open source crawling framework *scrapy*[7]. We configured the framework to work in a breadth-first-search-fashion way, and restricted the crawling the web pages belonging to the 32 selected web sites. Thereby we discard all other discovered web pages.

The obtained web site-specific crawl corpus contains more than 11.2 million HTML pages. The distribution of number of pages for each of the web sites is listed in Table 3. Although for some web sites the number of gathered pages looks comprehensive, we do not claim this crawl to be complete.

Together with the product-specific crawl corpus we provide over 13 million web pages retrieved from shopping web sites. The pages are provided within WARC files and can be downloaded from our web page.

---

[7] https://github.com/scrapy/scrapy.

Table 3. Number of pages per web site contained in the product data corpus

Web site	# Pages	Web Site	# Pages
target.com	2, 007, 121	frontierpc.com	187, 184
shop.com	1, 754, 368	abt.com	115, 539
walmart.com	1, 711, 517	flipkart.com	63, 112
selection.alibaba.com	607, 410	conns.com	61, 158
microcenter.com	459, 921	costco.com	54, 274
aliexpress.com	431, 005	dhgate.com	50, 099
ebay.com	413, 144	shop.lenovo.com	41, 465
macmall.com	401, 944	bjs.com	40, 930
apple.com	391, 539	vnewegg.com	37, 393
bestbuy.com	389, 146	microsoftstore.com	24, 163
techspot.com	386, 273	vsamsclub.com	22, 788
techforless.com	361, 234	tomtop.com	13, 306
overstock.com	347, 846	alibaba.com	7, 136
searsoutlet.com	341, 924	boostmobile.com	2, 487
pcrush.com	292, 904	sears.com	659
tesco.com	222, 802	membershipwireless.com	457

6   Baselines

In the following, we describe for each of the tasks of product feature extraction and product matching a set of straight-forward experiments and the resulting performance on the gold standard datasets described in Sects. 3 and 4. We performed these experiments in order to verify that the gold standards are challenging enough.

6.1   Product Feature Extraction Baseline

In order to create a baseline for the task of product feature extraction from web pages, we present a straight-forward approach and its results based on the feature extraction gold standard, presented in Sect. 3. For the evaluation we consider textual information from three different sources as input. The first source is information marked up with Microdata within the HTML page. As second source, we select product specification tables and as the third the specification lists.

6.1.1   Method

The approach makes use of the properties which are contained for the different products in the product catalog, described in Sect. 4. From these property names, we generate a dictionary, which we then apply to all web pages in the gold standard. This means, whenever the name of the feature within the catalog occurs on the web page, we extract this as feature for the product.

6.1.2   Results

We applied the dictionary method described above for the tree mentioned sources. The results for the dictionary approach vary for the different parts of the gold standard. However common for all results is underperformance of the method in general. Specifically, the method reaches results in the span of 0.400–0.600 F-score for all parts and all categories, meaning that improvement is needed. More closely, we can find that in general the method provides better recall (0.450–0.600) than precision (0.390–0.570). The reason for the poor performance can be found in the difference of the values coming from the product catalog and the different vendors. For instance, the size of a display in our catalog are inches, however some of the vendors use the metric system for that measure. Category wise, we can conclude that the *headphones* achieves the best results for all input sources, while *mobile phones* and *TVs* have comparable results.

6.2   Product Matching Baselines

In the following we present three different matching approaches and their results based on the product matching gold standard, presented in Sect. 4.

We use 3 distinct methodologies for feature extraction for the creation of the baselines: (i) bag-of-words (BOW), (ii) dictionary approach and (iii) text embeddings. For all the three approaches we consider textual information from three different sources as input. The first source is the HTML page itself, where we remove all HTML tags which are unrelated to the specification of the product. The second source of features are information marked up with Microdata within the HTML page. As third source, we select product specification tables and lists.

We take into account the textual information of the two input sources and preprocess the text by splitting it on non-alphanumeric characters. We convert all tokens to lower case and remove stopwords. Next, we apply a Porter Stemming Filter[8] to the remaining tokens. Finally, we take n-grams ($n \in 1, 2, 3$) of the resulting tokens.

Using the later described approaches we create vectors from the different input sources and compare them using three different similarities: string matching (sequential overlap of tokens), *Jaccard* similarity and *cosine* similarity based on TF-IDF vectors.

In the following we briefly explain each baseline method and discuss the best. Detailed results for each parameter settings can be found on our website.

---

[8] http://nlp.stanford.edu/IR-book/html/htmledition/stemming-and-lemmatization -1.html.

6.2.1   Bag-of-Words

The bag-of-words model is a simplifying representation where all tokens are used which are created from the preprocessing disregarding word order but keeping the multiplicity.

With this method we were able to reach 0.588, 0.412, and 0.552 F-score, for headphones, phones and TVs respectively. Generally, the precision and recall show equal performance, with the phones category being the exception. The results indicate that a purely BOW-based approach is not suitable for this task.

6.2.2   Dictionary

Similarly, like the feature extraction baseline shown in Sect. 6.1 we build the dictionary for the known attributes of the product catalog. Conversely, for each known attribute we construct a list of available attribute values. Subsequently, we tag potential values from the labeled set with the attributes from our dictionary.

With this method we were able to reach 0.418, 0.614, and 0.553 F-score, for headphones, phones and TVs respectively. As with the BOW approach, precision and recall have equal performance. Noteworthy is that the results for the dictionary are comparable to the BOW approach. This can be explained by the difference in values used by various web sites (see Sect. 6.1).

6.2.3   Paragraph2Vec

The most prominent neural language model for text embedding on a document level is *paragrph2vec* [5]. Paragraph2vec relies on two algorithms: Distributed Memory (DM) and Distributed Bag-of-Words (DBOW). For the purposes of this experiment we built a DBOW model using 500 latent features. To be able to represent document embeddings paragrapah2vec maps each document to a unique paragraph vector. DBOW ignores the context words in the input and instead forms a classification task given the paragraph vector and randomly selected words from a text window sample.

Paragraph2vec is able to reach 0.675, 0.613, and 0.572 F-score, for headphones, phones and TVs respectively. As expected, text embeddings outperform both BOW and the Dictionary approach.

7   Related Work

This section gives an overview of related research in the areas of product feature extraction and product matching and discusses evaluation datasets for these tasks.

Product Data Corpora. The most widely known public gold standard for product matching to this date is introduced in Köpcke et al. [3]. The evaluation datasets are based on data from Amazon-GoogleProducts and Abt-Buy[9]. However, the dataset contains only four attributes: name, description, manufacturer

---

[9] http://dbs.uni-leipzig.de/en/research/projects/object_matching/fever/
benchmark_datasets_for_entity_resolution.

and price. A more recent gold standard is introduced in Ristoski and Mika [14], where the authors provide a dataset marked up in Microdata markup from several web sites. The evaluation dataset was gathered was a subset from the Web-DataCommons structured dataset[10] and is gathered from several e-shops. However, the gold standard uses only two textual features: name and description. Besides, crawls from e-commerce web sites have also been published occasionally, like the one used in [6]. Unfortunately the data of such corpora mostly originates from one website, and is therefore not useful for identity resolution or data fusion. Furthermore, the data used by [9] which originated from different web site cannot be directly used for product matching as the authors did not focus on an overlap of products and therefore the usability for identity resolution is unclear.

Feature Extraction Methods. One of the most prominent studies for product feature extraction is Nguyen et al. [10]. The authors introduce a pipeline for product feature extraction and schema alignment on product offers from multiple vendors in order to build a product catalog. In [13] the authors use the Bing Crawl to extract features from HTML table and list specifications and showcase their system with a product matching use case. In order to identify HTML tables and lists on product web pages, they use several consecutive classification approaches, which we also use in order to identify the location of tables and lists on the web page. Again the used dataset is not publicly available, although the authors provide (some) of their results to the public. For the purposes of this study we have reimplemented the methodology for extracting feature-value pairs used in this study and we reached 0.724 F-score for tables and 0.583 F-score for lists.

In our previous works [11,12], we showed the usability of product-related Microdata annotations for product feature extraction. In particular the works underline that it is possible learning product-category-specific regular expressions to extract features particular from titles and descriptions of the products.

The work by Ristoski and Mika [14] uses the Yahoo product data ads to train *conditional random fields* for extracting product features from the titles as well as the descriptions product offers that were annotatated using the Microdata syntax. A similar work that employs conditional random fields for chunking product offer titles is [7].

Product Matching Methods. Recent approaches by [2] match unstructured product offers retrieved from web pages to structured product specification using data found in the Microsoft Bing Product catalog. A work focusing on the exploitation of product specific identifiers, like the *manufacturer part number* (MPN) or the GTIN for product matching is presented in [4]. In [1] the authors introduce a novel approach for product matching by enriching product titles with essential missing tokens and calculate the importance score computation that takes context into account.

---

[10] http://webdatacommons.org/structureddata/index.html.

All those works make use of proprietary data for the task of product matching, which on the one hand side makes it hard to validate their results. On the other hand side it is also not possible to compare results of different approaches, as the heavily depend on the used data.

All of the artifacts and results from this paper are available for download at http://webdatacommons.org/productcorpus/.

References

1. Gopalakrishnan, V., Iyengar, S.P., Madaan, A., Rastogi, R., Sengamedu, S.: Matching product titles using web-based enrichment. In: Proceedings of the 21st ACM International Conference on Information and Knowledge Management, CIKM 2012, pp. 605–614. ACM, New York (2012)

2. Kannan, A., Givoni, I.E., Agrawal, R., Fuxman, A.: Matching unstructured product offers to structured product specifications. In: 17th ACM SIGKDD International Conference On Knowledge Discovery and Data Mining, pp. 404–412 (2011)

3. Köpcke, H., Thor, A., Rahm, E.: Evaluation of entity resolution approaches on real-world match problems. Proc. VLDB Endowment 3(1–2), 484–493 (2010)

4. Köpcke, H., Thor, A., Thomas, S., Rahm, E.: Tailoring entity resolution for matching product offers. In: Proceedings of the 15th International Conference on Extending Database Technology, pp. 545–550. ACM (2012)

5. Le, Q.V., Mikolov, T.: Distributed representations of sentences, documents. arXiv preprint arXiv:1405.4053 (2014)

6. McAuley, J., Targett, C., Shi, Q., van den Hengel, A.: Image-based recommendations on styles and substitutes. In: Proceedings of the 38th International ACM SIGIR Conference on Research and Development in Information Retrieval, pp. 43–52. ACM (2015)

7. Melli, G.: Shallow semantic parsing of product offering titles (for better automatic hyperlink insertion). In: Proceedings of the 20th ACM SIGKDD International Conference on Knowledge Discovery and Data Mining, KDD 2014, pp. 1670–1678. ACM, New York (2014)

8. Meusel, R., Petrovski, P., Bizer, C.: The webdatacommons microdata, RDFa and microformat dataset series. In: Mika, P., Tudorache, T., Bernstein, A., Welty, C., Knoblock, C., Vrandečić, D., Groth, P., Noy, N., Janowicz, K., Goble, C. (eds.) ISWC 2014. LNCS, vol. 8796, pp. 277–292. Springer, Heidelberg (2014). doi:10.1007/978-3-319-11964-9_18

9. Meusel, R., Primpeli, A., Meilicke, C., Paulheim, H., Bizer, C.: Exploiting microdata annotations to consistently categorize product offers at web scale. In: Stuckenschmidt, H., Jannach, D. (eds.) EC-Web 2015. LNBIP, vol. 239, pp. 83–99. Springer, Heidelberg (2015). doi:10.1007/978-3-319-27729-5_7

10. Nguyen, H., Fuxman, A., Paparizos, S., Freire, J., Agrawal, R.: Synthesizing products for online catalogs. Proc. VLDB Endowment 4(7), 409–418 (2011)

11. Petrovski, P., Bryl, V., Bizer, C.: Integrating product data from websites offering microdata markup. In: Proceedings of the Companion Publication of the 23rd International Conference on World Wide Web Companion, pp. 1299–1304. International World Wide Web Conferences Steering Committee (2014)

12. Petrovski, P., Bryl, V., Bizer, C.: Learning regular expressions for the extraction of product attributes from e-commerce microdata (2014)
13. Qiu, D., Barbosa, L., Dong, X.L., Shen, Y., Srivastava, D.: Dexter: large-scale discovery and extraction of product specifications on the web. Proc. VLDB Endowment **8**(13), 2194–2205 (2015)
14. Ristoski, P., Mika, P.: Enriching product ads with metadata from HTML annotations. In: Proceedings of the 13th Extended Semantic Web Conference (2015, to appear)

MFI-TransSW+: Efficiently Mining Frequent Itemsets in Clickstreams

Franklin A. de Amorim[1]([✉]), Bernardo Pereira Nunes[1,3]([✉]), Giseli Rabello Lopes[2], and Marco A. Casanova[1]

[1] Department of Informatics, PUC-Rio, Rio de Janeiro, RJ, Brazil
{famorim,bnunes,casanova}@inf.puc-rio.br
[2] Department of Computer Science, UFRJ, Rio de Janeiro, RJ, Brazil
giseli@dcc.ufrj.br
[3] Department of Applied Informatics, UNIRIO, Rio de Janeiro, RJ, Brazil
bernardo.nunes@uniriotec.br

Abstract. Data stream mining is the process of extracting knowledge from massive real-time sequence of data items arriving at a very high data rate. It has several practical applications, such as user behavior analysis, software testing and market research. However, the large amount of data generated may offer challenges to process and analyze data at nearly real time. In this paper, we first present the MFI-TransSW+ algorithm, an optimized version of MFI-TransSW algorithm that efficiently processes clickstreams, that is, data streams where the data items are the pages of a Web site. Then, we outline the implementation of a news articles recommender system, called ClickRec, to demonstrate the efficiency and applicability of the proposed algorithm. Finally, we describe experiments, conducted with real world data, which show that MFI-TransSW+ outperforms the original algorithm, being up to two orders of magnitude faster when processing clickstreams.

Keywords: Datastream · Frequent itemsets · Data mining

1 Introduction

Data stream mining is the process of extracting knowledge from massive real-time sequence of possibly unbounded and typically non-stationary data items arriving at a very high data rate [2]. The mining of frequent itemsets in data streams has several practical applications, such as user behavior analysis, software testing and market research. Nevertheless, the massive amount of data generated may pose an obstacle to processing them in real time and, hence, to their analysis and decision making. Thus, improvements in the efficiency of the algorithms used for these purposes may bring substantial benefits to the systems that depend on them.

The problem of mining frequent itemsets in data streams can be defined as follows. Let $I = \{x_1, x_2, ..., x_n\}$ be a set of items. A subset X of I is called an itemset. A transactional data stream is a sequence of transactions,

© Springer International Publishing AG 2017
D. Bridge and H. Stuckenschmidt (Eds.): EC-Web 2016, LNBIP 278, pp. 87–99, 2017.
DOI: 10.1007/978-3-319-53676-7_7

$D = (T_1, T_2, ..., T_N)$, where a transaction T_i is a set of items and N is the total number of transactions in D. The number of transactions in D that contains X is called the support of X and denoted $sup(X)$. An itemset X is frequent iff $sup(X) \geq N \cdot s$, where $s \in [0, 1]$ is a threshold, defined by the user, called the minimum support. The challenge of mining frequent itemsets in data streams is directly associated with the combinatorial explosion of the number of itemsets and the maximum memory space needed to mine them. Indeed, if the set of items I has n elements, the number of possible itemsets is equal to $2^n - 1$. As the length of a data stream approaches a very large number, the probability of items to be frequent becomes larger and harder to control with limited memory.

A *clickstream* over a Web site is a special type of data stream where the transactions are the users that access the Web site and the items are the pages of the Web site.

MFI-TransSW [8] is an efficient algorithm, in terms of processing and memory consumption, for finding frequent itemsets in sliding windows over data streams, using bit vectors and bit operations. However, the lack of a more efficient sliding window step and the fact that MFI-TransSW algorithm was designed to process a finite sequence of transactions opens up new opportunities for improvement. In this paper, we introduce the MFI-TransSW+ algorithm, an optimized version of the MFI-TransSW to process clickstreams in real time. Additionally, we briefly describe ClickRec, an implementation of a news articles recommender system based on MFI-TransSW+. Finally, we present experiments, using real data, to evaluate the performance for clickstreams of the original MFI-TransSW and the proposed MFI-TransSW+ algorithm, as well as to validate the recommendations obtained with ClickRec.

The key contributions of the paper are twofold: a new strategy to maintain bit vectors, implemented in the MFI-TransSW+ algorithm, which leads to a substantial performance improvement when processing real-world clickstreams; a new recommendation strategy for news articles, based on measuring metadata similarity, which proved to double the recommendation conversion rate, when compared with the baseline recommendation algorithm adopted by a news portal.

The remainder of the paper is organized as follows. Section 2 reviews the literature. Section 3 presents the MFI-TransSW and the MFI-TransSW+ algorithms. Section 4 overviews the ClickRec recommender system and its implementation. Section 5 describes the experiments. Finally, Sect. 6 concludes the paper.

2   Related Work

The algorithms that mine frequent itemsets may be classified into those that adopt fixed windows and those that use sliding windows [4]. They may also be classified according to the results they produce into exact and approximate algorithms. The approximate algorithms may be further sub-divided into false positives and false negatives.

Agrawal et al. [1] proposed the A-Priori algorithm, one of the most important contributions to the problem of finding frequent itemsets. The A-Priori algorithm is based on the observation that, if a set of items i is frequent, then all subsets of i are also frequent.

Manku and Motwani [11] developed two single-pass algorithms, called Sticky-Sampling and Lossy Counting, to mine frequent itemsets in fixed windows. They also developed a method, called Buffer-Trie-SetGen (BTS), to mine frequent itemsets in data streams. Both algorithms have low memory usage, but they generate approximate results, with false positives.

Li et al. [9] proposed a single-pass algorithm, called DSM-FI, and Li et al. [10] defined the DSM-MFI algorithm to extract all frequent itemsets (FI) and maximal frequent itemsets (MFI) from a data stream. These algorithms use a data structure based on prefix-trees to store the items as well as auxiliary data.

Yu et al. [13] proposed false positive and false negative algorithms based on the Chernoff Bound [5] to mine frequent itemsets from high speed data streams. The algorithms use two parameters to eliminate itemsets and control memory usage.

Lee et al. [7] proposed a filtering algorithm to incrementally mine frequent itemsets in a sliding window. Chang et al. [3] proposed an algorithm based on BTS, called SWFI-stream, to find frequent itemsets in a transaction-sensitive sliding window. Chi et al. [6] proposed an algorithm, called Moment, to process data streams and extract closed frequent itemsets inside a transaction-sensitive sliding window. Li et al. [8] developed the MFI-TransSW algorithm to mine frequent itemsets in sliding windows over data streams, using bit vectors and bit operations.

The MFI-TransSW+ algorithm, proposed in this paper, processes click-streams in sliding windows. It adopts a version of the A-Priori algorithm to find all frequent itemsets, without generating false positives. Lastly, it maintains a list of users of the current window and circularly updates the bit vectors when the window moves, which result in expressive performance gains, when processing clickstreams, as discussed in Sect. 5.

3 The MFI-TransSW and the MFI-TransSW+ Algorithms

3.1 The MFI-TransSW Algorithm

The MFI-TransSW algorithm mines frequent itemsets in data streams. MFI-TransSW is a *single pass mining* algorithm in the sense that it reads the data only once and uses bit vectors to process the data that fall inside a sliding window and to extract the frequent itemsets, without generating false positives or false negatives.

Each item X that occurs in a transaction in the sliding window is represented as a bit vector Bit_X, where each position $Bit_X(i)$ represents the ith active transaction in the window and is set to 1 if the transaction includes X. For example, suppose that the window has size $W = 3$ and that the 3 active transactions are $T_1 = \{a, b\}$, $T_2 = \{b\}$ and $T_3 = \{a, b, c\}$. Then, the bit vectors corresponding to the items in the window W would be $Bit_a = 101$, $Bit_b = 111$ e $Bit_c = 001$. MFI-TransSW has three main stages, described in what follows: (1) *window initialization*; (2) *window sliding*; and (3) *frequent itemsets generation*.

The first stage processes the transactions in the order they arrive. For each transaction T and each item X in T, the initialization stage creates a bit vector Bit_X for X, if it does not exist, initialized with 0's. The bit corresponding to T in Bit_X is set to 1.

When the number of transactions reaches the size of the window, the window becomes *full* and the second stage starts. The sequence of transactions is treated as a queue, in the sense that the first and oldest transaction is discarded, opening space for the new transaction, which is added at the end of the queue. Let O be the oldest transaction and T be the new transaction. The bit vectors are updated accordingly: a *bitwise left shift* operation is applied to each bit vector, which eliminates the first bit, that is, the information about the oldest transaction O. The information about the new transaction T is registered at the end of the bit vectors: for each bit vector Bit_X, the last bit of Bit_X is set to 1, if T contains X, and set to 0, otherwise. If T contains an item for which no bit vector exists, a new bit vector is created as in the initialization stage. At this point, a cleaning operation eliminates all bit vectors that contain only 0's. This process continues as long as new transactions arrive or the stream is interrupted.

The third stage is an adaptation of the A-Priori algorithm [1] using bit vectors and may be executed at any time at the user request or whenever it becomes necessary to generate the frequent itemsets. This stage successively computes the set FI_k of frequent itemsets of size k. FI_1 contains all items X such that $sup(X)$, the *support* of X, defined as the number of bits set to 1 in Bit_X, is greater than a minimum threshold. CI_2 contains all pairs X, Y of frequent items in FI_1 and FI_2 contains all pairs X, Y in CI_2 such that $Bit_{XY} = Bit_X \wedge Bit_Y$ has at least as many bits set to 1 as the minimum threshold. This process continues to generate CI_k and FI_k until no more frequent itemsets are found.

One of the advantages of this approach is the low memory usage. Each item is associated with a bit vector whose length is the window size; hence, the total memory required in bytes is $((i \times w)/8)$, where i is the number of items and w is the number of transactions in the window. For example, a window with 100K transactions and 1K distinct items will require less than 12 MB of memory.

3.2   The MFI-TransSW+ Algorithm

In this section we describe the version of MFI-TransSW we propose, called MFI-TransSW+, which is optimized to process clickstreams. Our experiments indicated that MFI-TransSW+ is 2 orders of magnitude fasters than MFI-TransSW when processing clickstreams (see Sect. 5).

3.2.1 Processing Clickstreams

A *clickstream* over a Web site is a special type of data stream, where the transactions are the users that access the Web site and the items are the pages of the Web site [12]. The records of a clickstream usually have the fields: (1) *Timestamp*, the time the click was executed; (2) *User*, the user ID that executed the click; (3) *Page*, the page accessed; and (4) *Action*, the action executed by the user, such as a page access, vote, etc. (since we consider only page accesses, we will omit this field).

The original MFI-TransSW algorithm was designed to process a sequence of transactions, where each transaction is modeled as a finite set of well-know items. However, in a click stream, each item is a click that represents a pageview. Furthermore, we typically do not know when a transaction starts or ends, since each click record only indicates that a user accessed a page at a given time, and there is no action indicating that a user session started or ended.

Recall that, when the window moves and a new transaction arrives, the MFI-TransSW algorithm applies the shift left operation to all bit vectors, including the bit vectors that do not correspond to the items of the new transaction. However, in a clickstream, the set of pages a user accesses is typically a small fraction of the pages all users access in the window (see Sect. 5). Hence, the way MFI-TransSW updates bit vectors becomes inefficient. MFI-TransSW+ therefore adopts new data structures to optimize processing the bit vectors, as discussed in what follows.

From now on we refer to users, pages and pagesets, rather than transactions, items and itemsets, respectively.

3.2.2 Data Structures

We assume that users have a unique ID (*UID*) and that the sliding window has size w. Then, all bit vectors will also have size w. We introduce two new data structures. The *list of UID's*, *LUID*, is a circular list, with maximum size w, that contains the *UID's* of all users that occur in the current window, in the order they arrive. The position of each user in *LUID* indicates its position in the bit vectors. The *list of bit vectors per user*, *LBVPU*, stores, for each user u, the IDs of the bit vectors that have the bit corresponding to u set to 1.

These data structures are updated as follows. Suppose that the next record of the clickstream indicates that user u accessed page p.

Case 1. User u is in *LUID*, that is, u.has already been observed in the current window.

Case 1.1. Page p has been accessed before.
Then, there is a bit vector Bit_p already allocated to page p. We set $Bit_p(j) = 1$, where j is the position of u in *LUID*, and add the ID of Bit_p to *LBVPU*, associated with user u, if not already in *LBVPU*.

Case 1.2. Page p has never been accessed before.

We allocate a bit vector Bit_p for page p, initialize Bit_p with 0's, set $Bit_p(j) = 1$, where j again is the position of u in $LUID$, and add the ID of Bit_p to $LBVPU$, associated with user u.

Case 2. User u is not in $LUID$, that is, u has not been observed in the current window.

Case 2.1. Assume that the current window is not full.
We add u to the end of $LUID$ and update the data structures as in Cases 1.1 and 1.2.

Case 2.2. Assume that the current window is full.
We do not apply a shift left operation to all bit vectors, but update each bit vector and $LUID$ as circular lists with the help of a pointer k to the last position used. That is, the next position to be used is $((k + 1) \bmod w)$. We first set to 0 position k of the bit vectors associated, in the $LBVPU$ list, with the old user at position k of $LUID$ and remove the old user and all its bit vector IDs from $LBVPU$. The new user u is added to $LUID$ in position k, replacing the user that occupied such position. Next, we update the data structures as in Cases 1.1 and 1.2 for the new user at position k of $LUID$. Finally, we eliminate all bit vectors that contain only 0's.

3.2.3   Pseudo-code of the MFI-TransSW+ Algorithm

Algorithm 1 outlines the pseudo-code of MFI-TransSW+.

Lines 1 to 12 just repeat how the data structures are updated. Lines 1 to 3 initialize the data structures. Lines 4 to 12 process the next record of the clickstream. Line 4 reads the next record from the clickstream CDS, setting user u and page p. Line 5 verifies if user u already exists in $LUID$. If this is the case, Line 6 updates the data structures as in Case 1. Lines 8 to 12 are executed if user u does not exist in $LUID$. Line 8 verifies if $LUID$ is not full. If $LUID$ is not full, Line 9 updates the data structures as in Case 2.1. If $LUID$ is full, Lines 11 and 12 are executed. Line 11 moves pointer k to the next position to be updated. Line 12 updates the data structures as in Case 2.2.

Lines 13 to 21 generate frequent pagesets for the current window as in the original algorithm. Line 13 generates all frequent pagesets of size $j = 1$; it counts the number of bits 1 of the bit vectors and stores in FI_1 those that have more bits set to 1 than the minimum support $(s \cdot w)$ for the current window. Lines 14 to 21 are executed for each possible size j of pagesets. Line 14 initializes the loop with $j = 2$ and increments j by 1 until no new frequent pagesets is found. Line 15 generates CI_j from the list of frequent pagesets of size $j - 1$, as in the original algorithm. To compute the support of the new candidates, Line 16 ANDs the bit vectors of the pages in each pageset in CI_j. Lines 17 to 21 identify which candidates are frequent pagesets. Line 17 initializes FI_j, the list of frequent pagesets of size j. Line 18 selects a candidate c_j from CI_j. Line 19 verifies if the support of c_j is greater than or equal to $(s \cdot w)$, the minimum support. If this is the case, Line 20 adds c_j to FI_j. Lastly, Line 21 adds all pagesets in FI_j to the set FI-$Output$ of all frequent pagesets.

Algorithm 1. The optimized MFI-TransSW+ algorithm.

---

input : CDS - a clickstream

        s - a user-defined minimum support, ranging in the interval [0, 1]

        w - a user-defined window size

output: $FI\text{-}Output$ - a set of frequent page-access-sets

  /* Initialize the data structures.   */

1 $k \leftarrow 0$;

2 $FI\text{-}Output \leftarrow \emptyset$;

3 $LUID, LBVPU \leftarrow NULL$;

  /* Process the next record of the clickstream.   */

4 **for** $(u, p) \in CDS$ **do**

5   **if** $u \in LUID$ **then**

6   Update the data structures for (u, p) as in Case 1;

7   **else**

8   **if** $LUID \neq FULL$ **then**

9   Update the data structures for (u, p) as in Case 2.1;

10   **else**

11           $k \leftarrow (k + 1) \bmod w$;

12   Update the data structures for (u, p) as in Case 2.2;

  /* Generate frequent pagesets, at user's request.   */

13 $FI_1 \leftarrow \{$ frequent pagesets of size 1 $\}$;

14 **for** $(j \leftarrow 2; FI_{j-1} \neq NULL; j{+}{+})$ **do**

15   Generate CI_j from FI_{j-1} ;

16   Execute *bitwise AND* to compute the support of all pagesets in CI_j;

17      $FI_j \leftarrow \emptyset$;

18   **for** $c_j \in CI_j$ **do**

19   **if** $|sup(c_j)| \geq w \cdot s$ **then**

20           $FI_j \leftarrow FI_j \bigcup \{c_j\}$;

21      $FI\text{-}Output \leftarrow FI\text{-}Output \bigcup FI_j$;

---

4   ClickRec Recommender System

In this section, we describe ClickRec, a news articles recommender system that processes clickstreams generated by users of a news portal and recommends news articles based on mining frequent pagesets using the MFI-TransSW+ algorithm (Sect. 3).

We tested ClickRec with three different recommendation approaches: (1) treat the news articles as the items; (2) treat the metadata (e.g. tags, semantic annotations and even editorial category) of the news articles as the items; (3) treat the metadata of the news articles as the items, but adopting a similarity metric over metadata to create recommendations. Due to space limitations, we discuss only the third approach, which produced the best results. Indeed, during our experiments (see Sect. 5), we identified that most of the users (55%–70%)

accessed only one news article (these users are commonly called *bounce users*), which limits the practical application of the first approach.

The third approach works as follows. First, frequent pagesets are mined in a window W. Then, when a user u accesses an article a (that is, a page), the approach compares the metadata of a with the metadata of the most frequent pagesets, searching for the most similar set F. Finally, the first n news articles described by metadata most similar to F are recommended to u.

For example, suppose that a user accesses a news article having as metadata the semantic annotations $S = \{<barcelona>, <neymar>, <messi>\}$. Assume that $F = \{<barcelona>, <neymar>, <messi>, <BBVA\_league>\}$ was identified, in the last window, as the frequent itemset most similar to S. Then, the recommendation approach searches for other news articles described by metadata most similar to F. Similarity metrics such as TF-IDF were applied for this task.

ClickRec was implemented in Python, using Redis, a key-value database, to store the bit vectors and other data structures. It has three main modules: *Data Streams Processor*, *Frequent Itemsets Miner* and *Recommender*.

The *Data Streams Processor* module continuously processes clickstreams, maintaining a window of users. Using stages 1 and 2 of MFI-TransSW+, it creates or updates the bit vectors, and related structures, of the users belonging to the selected window. This module receives as input the window size (described in number of users).

The *Frequent Pagesets Miner* module can be invoked whenever the transaction window becomes full or after regular time intervals. When this module is activated, the current window is processed to mine the frequent pagesets by the application of stage 3 of MFI-TransSW+, using as input the bit vectors generated by *Data Streams Processor* module. This module has two configurable parameters for each module execution (saved in a setup file): the minimum support value and the maximum size of the frequent itemsets. In the experiments (Sect. 5), frequent itemsets of size larger than 4 were very time consuming to be mined and did not improve the recommendation results.

The *Recommender* module uses frequent pagesets mined by *Frequent Pagesets Miner* module as input to generate recommendations of news articles for a user. This module is invoked whenever a user click on a news article (the current news article information is also an input for this module) and generates associated recommendations in real time. This module is responsible for implementing the recommendation approaches mentioned earlier in this section.

5   Experiments and Results

This section reports two experiments conducted with the MFI-TransSW+ algorithm. The first experiment compares the performance of the original MFI-TransSW and the proposed MFI-TransSW+ algorithm, whereas the second experiment validates the recommendations provided by ClickRec. A user tracking system was used to collect clickstreams from one of the largest news Web sites in Brazil. In total, 25M of page views were collected per day. The experiments and analysis were performed using one-hour time frames containing, on the average, 1M of page views.

The experiments were performed on a standard PC with 2.5 GHz Intel Core i5 Processor and 16 GB RAM, using the implementations described in previous sections.

5.1 Comparison Between MFI-TransSW and MFI-TransSW+

To compare the performances of the original MFI-TransSW and the optimized MFI-TransSW+, we used a one-hour time frame containing 1M of page views. The experiment was divided into two parts. In the first part, a number of users were processed up to the point the window reached its full size. In the second part of the experiment, the clickstreams were processed until reaching a total of 10k window slides. The performance of both algorithms was only measured in the second part, since both algorithms adopt different approaches. Table 1 shows the results for different window sizes and Fig. 1 shows a significant performance gain between both algorithms. The gain in performance of the MFI-TransSW+ grows linearly with the increase of the window size. For a window size of w = 10,000, MFI-TransSW+ is up to 900 times faster than the original algorithm, that is, 2 orders of magnitude faster.

Table 1. Runtime comparison between MFI-TransSW and MFI-TransSW+.

Window size	Runtime (s)	
	MFI-TransSW	MFI-TransSW+
1,000	41.45	0.40
2,000	136.73	0.63
3,000	272.23	0.95
4,000	395.54	1.17
5,000	533.10	1.28
6,000	761.30	1.59
7,000	996.09	1.91
8,000	1,295.16	2.07
9,000	1,484.10	2.22
10,000	1,928.76	2.36

Note that the substantial difference in performance between the algorithms occurs because MFI-TransSW+ performs much less operations than MFI-TransSW in each window move. For instance, in a window of size $w = 10,000$, we may have a total of 20k bit vectors. So, when the window reaches its full size, the original algorithm executes a bitwise shift for every bit vector, whereas the optimized algorithm only performs the clean-and-update operation.

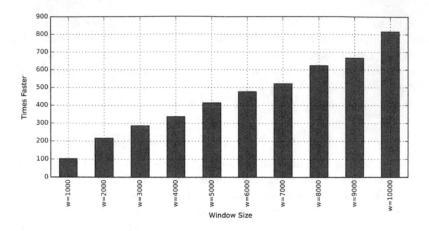

Fig. 1. Performance of the MFI-TransSW and MFI-TransSW+ algorithms.

5.2 Validation of ClickRec Recommendations

A preliminary analysis of the clickstreams over a period of one-hour indicated that the vast majority of users of the news portal under analysis were bounce users, that is, users that access only one page during its navigation. Figure 2 shows that the percentage of users that accessed only one page is much higher than those that accessed two or more pages. This analysis was conducted over two different editorial categories (A and B, as in Fig. 2) and shows similar user navigation behaviors.

The low number of pages accessed by users is a problem for frequent itemsets mining algorithms as there will be only a few to be mined. To solve this problem, instead of using user page views as input to our system, we opted to use the

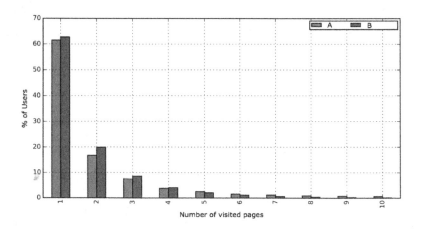

Fig. 2. Number of pages views per user.

third metadata approach described in Sect. 4 to mine the frequent itemsets. As mentioned in Sect. 4, all news articles have at least one semantic annotation, which were used in this experiment.

To run this experiment, we divided the clickstreams into pairs of two consecutive periods of one hour each. The first clickstream period is used to create the window and generate recommendations. The second clickstream period is used to extract a sample of users and measure the effectiveness of the recommendations. The sample consists only of users who have accessed more than one page during the second clickstream period.

Note that we need users who accessed more than one page because the first page accessed in the first period of the experiment will be used as input to the ClickRec recommendation module. Based on the first page accessed, ClickRec must recommend up to 10 news articles to a given user. If the user accessed one or more of the recommended pages, then the recommendation is considered successful.

Briefly, the following steps are performed for each pair of consecutive hours:

- **Loading:** Load a window with the first clickstream hour and generate the frequent itemsets. This step executes modules 1 and 2 of ClickRec.
- **Sampling:** Extract a sample of 10K users from the second clickstream hour.
- **News Recommendation:** Generate, for each user, recommendations according to his page accessed in the first clickstream period. This step executes module 3 of ClickRec.
- **Validation:** Validate if the user accessed a page recommended to him. If the user accesses the recommended page, we consider that the recommendation approach succeeded.

Figures 3 and 4 show the results for two distinct editorial categories (A and B). In the worst scenario, ClickRec was able to make correct recommendations for

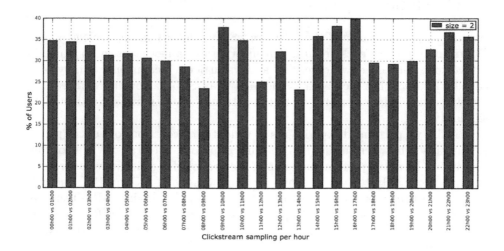

Fig. 3. ClickRec: number of correct recommendations for the A editorial category.

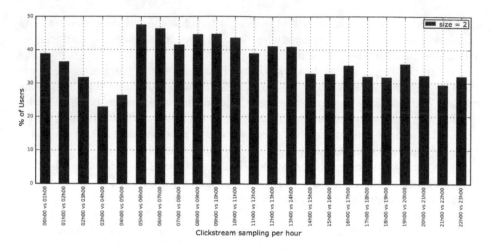

Fig. 4. ClickRec: number of correct recommendations for the B editorial category.

20% of the users. Although 20% seems to be low, the baseline algorithm provided by the news portal achieved a maximum conversion rate of 10% of the users that actually clicked in the recommended articles (unfortunately, the method and the portal name cannot be disclosed).

We also need to take into account the difficulty in measuring the correct recommendations. Indeed, a user may not access a recommended page either because it is irrelevant or because the page recommendation is not easily recognized due to design issues of the portal. Moreover, specifically in news portals, users tend to access only the top news articles, overlooking related articles.

6   Conclusions

In this paper, we presented the MFI-TransSW+ algorithm, an efficient mining algorithm to process clickstreams, that incorporates a new strategy to maintain bit vectors. Moreover, we described a news articles recommendation system, called ClickRec, that incorporates the MFI-TransSW+ algorithm and features a new recommendation strategy for news articles, based on measuring metadata similarity.

Finally, we performed experiments to validate the contributions of the paper. A comparison between the original and the proposed MFI-TransSW+ algorithm showed a substantial performance gain of the proposed algorithm, when processing clickstreams. In the experiments, for window sizes of 1K and 10K, the proposed algorithm was 100 to 900 times faster than the original MFI-TransSW algorithm. Furthermore, in the experiments with real data, ClickRec obtained much better conversion rates than the baseline algorithm currently adopted by the news portal.

References

1. Agrawal, R., Srikant, R.: Fast algorithms for mining association rules. VLDB **1994**, 1–32 (1994)
2. Babcock, B., Babu, S., Datar, M., Motwani, R., Widom, J.: Models and issues in data stream systems. In: ACM PODS, p. 1 (2002)
3. Chang, J.H., Lee, W.S.: A sliding window method for finding recently frequent itemsets over online data streams. J. Inf. Sci. Eng. **20**(4), 753–762 (2004)
4. Cheng, J., Ke, Y., Ng, W.: A survey on algorithms for mining frequent itemsets over data streams. Knowl. Inf. Syst. **16**(1), 1–27 (2008)
5. Chernoff, H.: A measure of asymptotic efficiency for tests of a hypothesis based on the sum of observations. Ann. Math. Stat. **23**, 493–507 (1952)
6. Chi, Y., Wang, H., Philip, S.Y., Muntz, R.R.: Catch the moment: maintaining closed frequent itemsets over a data stream sliding window. Knowl. Inf. Syst. **10**(3), 265–294 (2006)
7. Lee, C.-H., Lin, C.-R., Chen, M.-S.: Sliding window filtering: an efficient method for incremental mining on a time-variant database. Inf. Syst. **30**(3), 227–244 (2005)
8. Li, H.-F., Lee, S.-Y.: Mining frequent itemsets over data streams using efficient window sliding techniques. Expert Syst. Appl. **36**(2), 1466–1477 (2009)
9. Li, H.-F., Lee, S.-Y., Shan, M.-K.: An efficient algorithm for mining frequent itemsets over the entire history of data streams. In: Proceedings of the First International Workshop on Knowledge Discovery in Data Streams (2004)
10. Li, H.-F., Lee, S.-Y., Shan, M.-K.: Online mining (recently) maximal frequent itemsets over data streams. In: RIDE-SDMA, pp. 11–18. IEEE (2005)
11. Manku, G.S., Motwani, R.: Approximate frequency counts over data streams. VLDB **2002**, 346–357 (2002)
12. Montgomery, A.L., Li, S., Srinivasan, K., Liechty, J.C.: Modeling online browsing and path analysis using clickstream data. Mark. Sci. **23**(4), 579–595 (2004)
13. Yu, J.X., Chong, Z., Lu, H., Zhang, Z., Zhou, A.: A false negative approach to mining frequent itemsets from high speed transactional data streams. Inf. Sci. **176**(14), 1986–2015 (2006)

Reranking Strategies Based on Fine-Grained Business User Events Benchmarked on a Large E-commerce Data Set

Yang Jiao[1(✉)], Bruno Goutorbe[2], Matthieu Cornec[2], Jeremie Jakubowicz[1],
Christelle Grauer[2], Sebastien Romano[2], and Maxime Danini[2]

[1] SAMOVAR, CNRS, Telecom SudParis, Univ. Paris-Saclay,
9 Rue Charles Fourrier, 91011 Evry, France
{yang.jiao,jeremie.jakubowicz}@telecom-sudparis.eu
[2] Cdiscount, 120 Quai Bacalan, Bordeaux, France
{bruno.goutorbe,matthieu.cornec,christelle.grauer,sebastien.romano,
maxime.danini}@cdiscount.com

Abstract. As traditional search engines based on the text content often fail to efficiently display the products that the customers really desire, web companies commonly resort to reranking techniques in order to improve the products' relevance given a user query. For that matter, one may take advantage of fine-grained past user events it is now feasible to collect and process, such as the clicks, add-to-basket or purchases. We use a real-world data set of such events collected over a five-month period on a leading e-commerce company in order to benchmark reranking algorithms. A simple strategy consists in reordering products according to the clicks they gather. We also propose a more sophisticated method, based on an autoregressive model to predict the number of purchases from past events. Since we work with retail data, we assert that the most relevant and objective performance metric is the percent revenue generated by the top reranked products, rather than subjective criteria based on relevance scores assigned manually. By evaluating in this way the algorithms against our database of purchase events, we find that the top four products displayed by a state-of-the-art search engine capture on average about 25% of the revenue; reordering products according to the clicks they gather increases this percentage to about 48%; the autoregressive method reaches approximately 55%. An analysis of the coefficients of the autoregressive model shows that the past user events lose most of their predicting power after 2–3 days.

Keywords: Product search · Reranking · Autoregressive model

1 Introduction

The recent growth of on-line retail industry has made on-site product search engine a salient part of e-commerce companies. Product search is not only a problem of significant commercial importance, it also raises fundamental research

© Springer International Publishing AG 2017
D. Bridge and H. Stuckenschmidt (Eds.): EC-Web 2016, LNBIP 278, pp. 100–110, 2017.
DOI: 10.1007/978-3-319-53676-7_8

questions at the intersection of natural language processing, machine learning and information retrieval. The catalog of products of the largest companies can reach millions – if not tens of millions – of items, while user queries are typically made of very few words carrying limited semantic content. This greatly hampers the performance of traditional search engines based on text retrieval, in terms of conversion of the displayed results to purchases. Many companies thus opt for strategies to rerank the products using additional sources of information, in order to achieve better user satisfaction and larger revenue.

Fortunately, sophisticated tracking systems and 'big data' technologies now make it feasible to collect, store and process all user paths of the form:

$$\text{query} \rightarrow \text{click on product} \rightarrow \text{add-to-basket} \rightarrow \text{purchase},$$

over the whole site. It is then straightforward to build indicators with a granularity at the product level following a user query: e.g., number of clicks, add-to-basket and purchases per date. These numbers can directly serve the reranking purpose, if one argues that relevant products are simply those most likely to be viewed or purchased. This purely user behavior-based point of view leads to simple and objective reranking strategies, but it is not exempt from criticism. For instance, some products (such as erotic items) are likely to attract many curiosity clicks, and could therefore end up polluting many otherwise unrelated queries. Nevertheless, we believe that the use of past user events has the potential to improve conversion rates on e-commerce websites.

Previous studies discussed reranking strategies based on click data to improve retrieval of relevant web documents (Joachims et al. 2005; Agichtein et al. 2006) or images (Jain and Varma 2011). Jiao et al. (2015) exploited purchase data to improve product search performance via a collaborative filtering framework. In the present work we had access to a real-world data set of click, add-to-basket and purchase events collected over a five-month period from Cdiscount, a leading French e-commerce company. Based on this, our objective is to quantify the improvements brought by reranking strategies on top of a state-of-the-art semantic search engine using the BM25 statistics (Robertson et al. 1995). The most straightforward strategy consists in re-ordering products according to the clicks they gather over a fixed period after a user query: this follows the philosophy of previous works, applied to different contexts (Joachims et al. 2005; Agichtein et al. 2006; Jain and Varma 2011). We further propose a more sophisticated method that combines the three aforementioned types of event within an autoregressive model to predict the number of purchases. To the best of our knowledge, this work represents the first effort to benchmark reranking algorithms on real-world data set within an e-commerce context, and that exploits all the major types of implicit user feedback for that matter.

As for the performance metric on which the algorithms shall be evaluated, we believe that it is unnecessary for it to rely on subjective, human assigned relevance scores (Liu et al. 2007; Voorhees 2003) Since we work with retail data, we argue that the most relevant and objective performance metric is the average percent revenue generated by the top k displayed products, or *revenue@k*, which can be seen as a natural extension of the widely used *precision@k* Wiener 1956. The data large e-commerce companies have at their disposal are largely sufficient to compute meaningful estimates of *revenue@k*.

The rest of this paper is organized as follows. Section 2 describes the data set used in this study. Section 3 introduces reranking strategies, which include BM25 similarity Robertson et al. 1995, crude methods based on collected clicks, purchases or revenue, and our proposed autoregressive model fitted on all types of event. Section 4 deals with the evaluation metric, *revenue@k*. Section 5 gives the results and discussion related to the benchmarking of reranking methods on the data set.

2   Data Set

The raw data set was provided by Cdiscount, a leading online retailer in the French market: it consists of navigation logs and sale records over a period of 150 days from July 1st, 2015 to November 27, 2015, and contains several millions of distinct user queries per month. As can be expected, purchases are extremely unevenly distributed amongst the queries: Fig. 1a shows a long tail of queries concentrating a large number of purchases approximately following a power law. We focus this work on the top 1000 queries, which generate a significant part of all purchases made through the search engine (Fig. 1b).

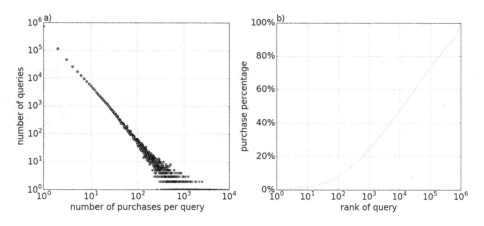

Fig. 1. (a) Distribution of the purchases by query; (b) cumulative percentage of purchases associated with queries (queries generating most purchases first).

The raw data contain a large amount of information related to user navigation through the website. We pruned and simplified the data structure in the following way. First, we only considered three types of event related to a product following a typed query: click, add-to-basket and purchase. "Negative" feedback events such as remove-from-basket or purchase abandon could also provide useful information, but we believe they would only marginally improve the reranking strategies. Second, we processed navigation sessions containing multiple searches by assigning the last typed query to each observed event: we thus obtained a list of more than 21 million pairs {query, product} labeled with a time stamp and an event tag: click, add-to-basket or purchase (see Table 1 for a summary of the final data set). It is then straightforward to count the number of each type of event associated with a given query and product at the desired temporal granularity: an example is given in Fig. 2.

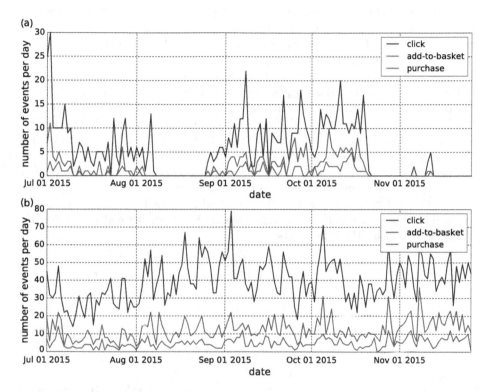

Fig. 2. Examples of times series of number of clicks, add-to-basket and purchases per day related to two different products following the user query 'printer'.

3   Reranking Strategies

Queries typed by users in e-commerce search engines are typically extremely short and carry little semantic content (Singh et al. 2012). Traditional engines looking for products whose description best match the queries' keywords often fail to display the relevant products, i.e. those most likely to be purchased, because there are too many matches. For example, it is difficult to distinguish semantically amongst *iPhone 4*, *iPhone 5* and *iPhone 6* with respect to the

Table 1. Summary of the final data used in this study

150	Days
1000	Distinct queries
746,884	Distinct products
21,450,377	User events[a]

[a]clicks, add-to-basket, purchases

query 'iPhone'. This is a particularly salient point, because customers are less and less inclined to crawl through pages of results until they reach the product they desire (Spink et al. 2002) and, given the number of e-commerce actors, may simply turn to a competitor.

Many web companies thus resort to *reranking* strategies, wherein additional information (such as popularity, price or sales) is integrated to reorder the displayed items. The ecommerce-specific data presented in Sect. 2 prove of valuable help for that matter: as mentioned in the introduction, one can follow previous works (Joachims et al. 2005; Agichtein et al. 2006; Jain and Varma 2011) and simply use clicks gathered by products. We also included for comparison a random and a BM25 (Robertson et al. 1995) ordering, and we finally implemented a more sophisticated method based on an autoregressive model.

Specifically, for a given user query, we started with the list of top 28 products returned by Cdiscount internal search engine. We then defined the following strategies to reorder these products:

1. *random reordering*, which any method should outperform.
2. *BM25*: BM25 (Robertson et al. 1995) is considered as a state-of-the-art similarity metric, so we used it as reference text-based ranking technique.
3. *reranking by click*: products are ordered by decreasing number of views collected after the query over a recent period. Previous works used user clicks to better retrieve relevant web documents or images (Joachims et al. 2005; Agichtein et al. 2006; Jain and Varma 2011), so it is natural to extend this approach to the e-commerce context.
4. *reranking by multivariate auto-regression*: products are ordered by decreasing revenue, deduced from an autoregressive model fitted on the three types of event described in Sect. 2. The model is described in detail in the next section.

3.1 Reranking by Multivariate Auto-Regression

Several points should be raised to justify the choice of a more sophisticated reranking model. First, since e-commerce business is more interested in maximizing the revenue than the number of views, the former quantity should guide the reranking strategy. However, the number of purchases is a weak signal (Fig. 2 and Table 1) so it is not optimal to use it as a sole predictor for the revenue; it is desirable to make use of the number of clicks and add-to-basket as well, as these signals are much stronger and highly correlated to the purchases (Fig. 2).

Finally, one may apply a temporal weighting scheme to penalize old signal rather than define a fixed period window; such a weighting scheme should ideally reflect the autocorrelation structure of the data, and not *a priori* subjective choices.

These considerations led us to select the vector autoregression (VAR) model (Hamilton 1994) in order to predict the number of purchases of a product following a given query from past time-series of clicks, add-to-basket and purchases. Specifically, we start with the following multivariate time-series:

$$
\mathbf{N}(t) = \begin{bmatrix} n_{\mathrm{c}}(t) \\ n_{\mathrm{a}}(t) \\ n_{\mathrm{p}}(t) \end{bmatrix},
\tag{1}
$$

where t represents the date and $n_{\mathrm{c}}(t), n_{\mathrm{a}}(t), n_{\mathrm{p}}(t)$ the number of clicks, add-to-basket and purchases related to some product after a query, respectively. The VAR model then describes the evolution of $\mathbf{N}(t)$ as a linear function of its past values:

$$
\mathbf{N}(t) = \sum_{i=1}^{P} \mathbf{A}_i \mathbf{N}(t-i) + \mathbf{E}(t),
\tag{2}
$$

where $\mathbf{A}_i$ is a time-invariant matrix containing the coefficients of the linear relation between the signal and itself for a time lag of i, and $\mathbf{E}(t)$ represents Gaussian noise. The maximum time lag, P, is called the order of the process.

The matrices $\mathbf{A}_i$ can be estimated from a least square fit on the observed signal including a ℓ_1 (or Lasso) regularization which seeks to minimize:

$$
\sum_{t} \left\| \mathbf{N}(t) - \sum_{i} \mathbf{A}_i \mathbf{N}(t-i) \right\|_2^2 + \lambda \sum_{i} \|\mathbf{A}_i\|_1 \cdot,
\tag{3}
$$

where $\|.\|_2$ denotes the Euclidean norm and $\|.\|_1$ the ℓ_1-norm. The parameter λ, which controls the regularization strength, is optimized using a three-fold cross-validation procedure, and takes typical values between 0.2 and 2. In the present work we estimated best-fitting matrices $\mathbf{A}_i$ for each pair $\{\text{query}, \text{product}\}$, but one may alternatively choose to aggregate signals by making one fit per query, or even one single global fit on all available pairs. This can be of interest in a production environment to save computational time and rerank queries associated with few events.

It is straightforward to use the VAR model for the reranking purpose. First, the products' number of purchases after a given query is predicted from past series from Eq. (2). The price is then taken into account to reorder the products associated with the query by decreasing predicted revenue.

3.2 Granger Causality Test

Before we proceed with the evaluation metric, it is worth verifying whether the time series we included in the VAR model are really able to forecast future purchase. Granger causality test (Granger 1969) is a statistical hypothesis test that can help answer this question. A time series $X(t)$ is said to Granger-cause $Y(t)$ if $Y(t)$ can be better predicted using the histories of both $X(t)$ and $Y(t)$ than it can by using the history of $Y(t)$ alone.

Explicitly speaking, one can test for the absence of Granger causality by estimating the following VAR model:

$$Y(t) = \alpha_0 + \sum_i \alpha_i Y(t-i) + \sum_i \beta_i X(t-i) + \mu(t) \tag{4}$$

We define the following null hypothesis:

$$H_0 : \beta_1 = \beta_2 = \cdots = \beta_p = 0. \tag{5}$$

A rejection of H_0 implies there is Granger causality.

We selected at random 1000 pairs of $\{query, product\}$ from our data set and performed Granger causality tests on both click and add-to-basket signals against purchase signal. The null hypothesis was rejected on over 98% of the pairs examined. Most of the unrejected cases correspond to less popular queries and products with negligible signals and accordingly large confidence intervals.

4   Evaluation Metric

Ranking algorithms are often evaluated with the *precision@k* metric, which is the average proportion of relevant items amongst the top k results. This metric relies on the availability of a labeled test data set wherein human annotators decide which items are relevant with respect to a user search (Liu et al. 2007; Voorhees 2003). However, there is no such data set publicly available in the e-commerce area, and manually assigning relevance scores on our data would be prohibitively time-consuming. Besides, human scores may suffer from severe inconsistencies, as annotators can have different relevance judgment with respect to the same item (Agrawal et al. 2009). Finally, relevance in the e-commerce context is highly volatile, as the desired products after a given query can vary from a period to another. For example, the relevance of the *iPhone 5* product with respect to the query 'iPhone' is likely to collapse when the *iPhone 6* is released.

We therefore argue that a better metric in the e-commerce context should be guided by the increase in revenue generated by the reranked items. After all, it is better to adopt a customer-oriented relevance score, directly related to the probability of purchase. Specifically, as a natural extension of the *precision@k* metric, we propose the *revenue@k* metric, which measures the percent revenue generated by the top k items of the search results, relative to the list of 28 products:

$$revenue@k = \frac{\sum_{\text{queries}} \text{revenue of the top } k \text{ products}}{\text{total revenue}}. \tag{6}$$

In the present work, the *revenue@k* metric serves to evaluate the reranking strategies outlined in Sect. 3.

5   Results and Discussion

We benchmarked the reranking strategies described in Sect. 3 over a 7-day test period, from November 21st to November 27th, 2015. The reranking-by–click strategy aggregated the number of click events over a 30-day interval ending at the day before each test date, and reordered accordingly the list of products associated with each query. The VAR model described in Sect. 3.1 was fitted on all available time series, i.e. number of clicks, add-to-basket and purchases per day, from the first available day (July 1st, 2015) until the day before each test date. This allowed making a prediction for the number of purchases at the test date and, together with the products' price, to reorder them by decreasing predicted revenue.

The algorithms were then evaluated using the *revenue@k* metric (Sect. 4), which calculates the percent revenue generated by the top k reranked products relative to the list of 28 products, averaged over the queries and over the testing dates, using the data set of purchase events (Sect. 2). Figure 3 and Table 2 show the performance of the reranking strategies according to that metric; they also display the upper limit of the revenue, which would be reached if one knew in advance the purchase events of the testing dates.

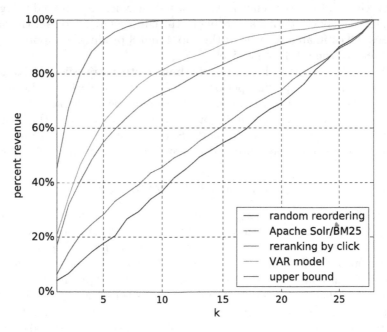

Fig. 3. Average percent revenue generated by the top k products (relative to the list of 28 products) of each reranking algorithm.

Table 2. Evaluation scores of the reranking algorithms according to the *revenue@k* metric.

	Estimated revenue generated by:	
	top 4 products	top 8 products
Random	14.6%	29.3%
Solr/BM25	24.8%	39.2%
Reranking by click[a]	48.4%	67.8%
VAR model	54.7%	76.1%
Upper limit	88.3%	98.8%

[a] aggregated over the 30 days preceding the test date

As is to be expected, the revenue is on average uniformly distributed amongst randomly ordered products, as shown by the linear trend in Fig. 3. A purely text-based strategy significantly improves the share of top ranked products: reordering the items according to their BM25 similarity (Robertson et al. 1995) with the query allows the top 4 and 8 products to increase their share of the estimated revenue by about 70% and 30%, respectively, compared to a random ordering (Table 2). Logically, products best matching the search terms are more likely to end up purchased. However, a much stronger enhancement is achieved using past user events: crudely reordering according to the clicks aggregated over the last 30 days raises the estimated revenue of the top 4 and 8 products to approximately 48% and 68% of the total, respectively.

The VAR model achieves better results than all other strategies. Although the leap is not as impressive as that performed by the simple, click-based algorithm compared to the BM25, the improvement should not be overlooked: it increases the estimated share of the top 4 and 8 products by about 12–13%, thus reaching 62% and 77% of the upper limit (Table 2). Such a gain is likely to translate into significantly larger conversion rate and revenue in most e-commerce companies.

5.1   Analysis of the Coefficients of the VAR Model

Before we conclude, it is interesting to analyze the coefficients of the VAR model in order to better understand how the different types of events are related to the number of purchases, and how their predicting power decay with time. We are interested in the elements of the third line of the matrices $\mathbf{A}_i$, which correspond to the coefficients of the linear relation between (1) the daily number of purchases and (2) the daily number of clicks, add-to-basket and purchases time-lagged by i (Eqs. 1 and 2). Figure 4 shows these coefficients, obtained by applying the VAR model to normalized time series, i.e. with zero mean and unit variance, and averaged over pairs of {query, product} and testing days. The normalization procedure ensures that the differences between the values of the coefficients do not reflect the systematic differences in amplitude between the time series (e.g., Fig. 2).

Interestingly, Fig. 4 suggests that clicks and (to a less extent) add-to-basket possess a stronger predictive power than purchases to forecast the latter time

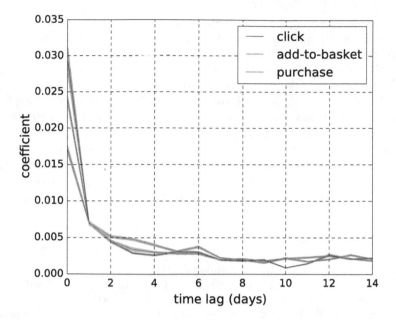

Fig. 4. Normalized coefficients of the VAR model related to the number of purchases, averaged over pairs of {query, product} and testing days, function of time lag. Shaded areas show one standard error around the mean.

series. We believe that this is explained by the larger amplitude of the former time series which, together with their strong correlation with the latter signal, allow to anticipate finer tendencies of the purchases to come: a close inspection of Fig. 2a reveals that clicks without purchases sometimes precede a wave of purchase events.

As can be expected, events lose their predicting power as time lag increases. The typical decay time may seem rather short: fitting an exponentially decaying function show that the 99% of the decrease of the average coefficient associated with the number of clicks is achieved in about 2.8 days. This value is approximately 3.3 days and 4.5 days for the add-to-basket and purchases events, respectively. So, purchases retain predictive power over a longer time, probably because they are not as volatile as other events.

6   Conclusion

For the first time in the e-commerce context (to the best of our knowledge), several reranking strategies were benchmarked on a real-world data set of user events provided by a leading online retail company. We used an evaluation metric adapted to e-commerce data, which measures the percent revenue generated by the top k items of the search results.

A text-based reordering according to the BM25 similarity (Robertson et al. 1995) between the products' description and the query's keywords allows the top four products to capture on average about 25% of the query's revenue. This is much less than reranking products according to the clicks they gathered following the query over the last 30 days, which increases this percentage to about 48%. A linear autoregressive method forecasting the number of purchases from past time-series of daily clicks, add-to-basket and purchases further reaches about 55%. The strength of the latter approach lies in its implicit exploiting of correlation of all major user events with purchases, and of their decay in predictive power as time lag increases.

The present work thus showed how crucial to the business model reranking algorithms guided by past user events are. As future lines of improvement, one may suggest the use of nonlinear predictive model and/or the integration of additional types of event such as remove-from-basket or purchase abandon.

References

Joachims, T., Granka, L., Pan, B., Hembrooke, H., Gay, G.: Accurately interpreting clickthrough data as implicit feedback. In: Proceedings of ACM Conference on Research and Development on Information Retrieval (SIGIR) (2005)

Agichtein, E., Brill, E., Dumais, S.: Improving web search ranking by incorporating user behavior information. In: Proceedings of ACM Conference on Research and Development on Information Retrieval (SIGIR) (2006)

Jain, V., Varma, M.: Learning to re-rank: query-dependent image re-ranking using click data. In: Proceedings of the 20th International Conference on World Wide Web (WWW). ACM (2011)

Jiao, Y., Cornec, M., Jakubowicz, J.: An entropy-based term weighting scheme and its application in e-commerce search engines. In: International Symposium on Web Algorithms (iSWAG) (2015)

Robertson, S.E., Walker, S., Jones, S., Hancock-Beaulieu, M.M., Gatford, M., et al.: Okapi at TREC-3. In: Proceedings of the Third Text REtrieval Conference (TREC-3) (1995)

Liu, T.-Y., Jun, X., Qin, T., Xiong, W., Li, H.: LETOR: benchmark dataset for research on learning to rank for information retrieval. In: Proceedings of SIGIR Workshop on Learning to Rank for Information Retrieval (2007)

Voorhees, E.M.: Overview of TREC. In: Proceedings of the Text REtrieval Conference (2003)

Wiener, N.: The theory of prediction. Modern Mathematics for Engineers (1956)

Singh, G., Parikh, N., Sundaresan, N.: Rewriting null e-commerce queries to recommend products. In: Proceedings of the 21st international conference on World Wide Web (WWW). ACM (2012)

Spink, A., Jansen, B.J., Wolfram, D., Saracevic, T.: From e-sex to e-commerce: web search changes. Computer **35**(3), 107–109 (2002). IEEE

Hamilton, J.D.: Time Series Analysis, vol. 2. Princeton University Press, Princeton (1994)

Granger, C.W.: Investigating causal relations by econometric models, cross-spectral methods. Econometrica: Journal of the Econometric Society, pp. 424–438 (1969)

Agrawal, R., Halverson, A., Kenthapadi, K., Mishra, N., Tsaparas, P.: Generating labels from clicks. In: Proceedings of ACM Conference on Web Search and Data Mining (WSDM) (2009)

Feature Selection Approaches to Fraud Detection in e-Payment Systems

Rafael Franca Lima$^{(\boxtimes)}$ and Adriano C.M. Pereira

Department of Computer Science (DCC),
Federal University of Minas Gerais (UFMG),
Belo Horizonte, Minas Gerais 31270-901, Brazil
{rafaelfrancalima,adrianoc}@dcc.ufmg.br

Abstract. Due to the large amount of data generated in electronic transactions, to find the best set of features is an essential task to identify frauds. Fraud detection is a specific application of anomaly detection, characterized by a large imbalance between the classes, which can be a detrimental factor for feature selection techniques. In this work we evaluate the behavior and impact of feature selection techniques to detect fraud in a Web Transaction scenario. To measure the effectiveness of the feature selection approach we use some state-of-the-art classification techniques to identify frauds, using real application data. Our results show that the imbalance between the classes reduces the effectiveness of feature selection and that resampling strategy applied in this task improves the final results. We achieve a very good performance, reducing the number of features and presenting financial gains of up to 57.5% compared to the actual scenario of the company.

Keywords: e-Commerce · Web · Fraud detection · e-Payment systems · Feature selection · Anomaly detection · Resampling

1 Introduction

People tend to use e-commerce systems because it is easy and convenient. However, this popularity generates a huge increase in the number of online frauds, resulting in billions of dollars losses each year worldwide. The works that deal with this topic usually try to find anomaly patterns that can be considered as a fraudulent behavior, such as a fraud in a Web transaction scenario [1–3].

Although we find several works in this research topic [4], there are still some points of improvement to detect frauds in electronic transactions. One of these points is related to how to filter the large amount of data generated in electronic transactions and transform this data in useful information. The main task in this context is to find a subset of features that are able to identify the anomalous behavior, which is called Feature Selection [5].

In general, the feature selection techniques perform a selection of features based on the class value. However, in anomaly detection scenario, as fraud

© Springer International Publishing AG 2017
D. Bridge and H. Stuckenschmidt (Eds.): EC-Web 2016, LNBIP 278, pp. 111–126, 2017.
DOI: 10.1007/978-3-319-53676-7_9

detection, the high imbalanced distribution between classes (e.g., distribution of fraud and non-fraud classes) generates new challenges to the feature selection techniques, which tend to select attributes in favor of the dominant class [6].

Thus, in this work we investigated how the imbalance between the classes affect the selection of attributes and, consequently, the models to detect fraud. To guide this study we enumerate some research hypotheses:

1. The high class imbalance reduces the effectiveness of feature selection to detect anomalies in electronic transactions;
2. Traditional methods of Feature Selection are not suitable to detect anomalies;
3. The reduction of class imbalance, using some resampling approaches, before the feature selection step, can improve the effectiveness of the selection of attributes.

To check these hypotheses, in this work we conduct experiments using three popular traditional feature selection techniques, i.e., techniques that do not use different strategies for the imbalance between the classes. We applied these techniques in two distinct distributions of the same dataset. The first one keeps the real proportion of fraud, while the second one reduces the imbalance between classes using some resampling methods. Thus, it is possible to measure how the imbalance between classes affects the selection of attributes.

To perform a deep investigation in resampling approaches, we evaluate 7 methods to reduce the class imbalance before feature selection step, including a resampling method created by us, specifically for this function. To build effectiveness models to identify frauds and evaluate the attributes selected by each approach, we used four classification techniques.

In order to test and validate these approaches, we conducted our experiments in an actual dataset from one of the most popular electronic payment systems in Latin America. In Sect. 5.1 we present a better characterization of this dataset. To evaluate the models we use three performance metrics related to classification accuracy (F1 and AUC) and Economic Efficiency (EE).

We should emphasize that in this work we use only this data set as a test case, since we did not find another similar one available. Despite this, we believe that our methodology is generic and can be applied to fraud detection in other Web scenarios.

The main contributions of this work are:

- The analysis of the effectiveness of traditional methods for feature selection in the detection of anomalies;
- A deep investigation in effectiveness of resampling methods before feature selection techniques;
- The creation of a resampling method to be used before feature selection step in anomalous scenario;
- The construction of a model for fraud detection, that combine strategies of resampling before feature selection and classification techniques;
- A complete validation of the proposed method using an actual data from an electronic payment system.

The remainder of this paper is organized as follows. Section 2 presents some related work. Section 3 describes the main fundamentals about resampling, feature selection and classification techniques. Section 4 shows the methodology used in this work. Section 5 describes our case study, using a representative sample of actual data, where we present a dataset overview and main results. Finally, Sect. 5 presents the conclusions and future work.

2   Related Work

Methods for detecting fraud in web-transactions are extensively discussed in the scientific community [2,4,7]. In [1] they are used the support vector machines (SVM), random forests and logistic regression to detect credit card fraud in a actual data set. In [8] it is used Networks Discovery methods and data mining techniques to detect fake users in a Service User Accounts of Web. In [3] the authors used pattern recognition techniques to detect products not delivered by fake sellers a in Free Market E-Commerce system.

Although the works have focus on different techniques to detect frauds, none of these works focus on feature selection to fraud detection. In those studies they were not using different approaches for feature selection to suit an anomaly detection application.

To confirm this lack we performed a systematic literature review among fraud detection works, following the methodology described in [9]. In this review we evaluate 30 most cited works and 20 most relevant researches in fraud detection. We do not identify any use of appropriate *feature selection* approaches or investigation in resampling methods before feature selection to identify frauds.

We found some researches in feature selection for high imbalanced dataset [6,10–13]. However, these works do not conduct fraud detection experiments. The main scenarios discussed in these researches are the biological genes (micro-array analysis) and diseases. Although, these scenarios also contain high imbalance between classes, the bases used differ in the ratio between classes, attributes, numbers and nature of the data.

There are works that discuss methods of resampling to reduce the high imbalance between classes on training for classification [14], but few studies have analyzed the application of these methods before the feature selection step. Resampling methods before feature selection step was used in [15,16]. However, these studies only exploit micro-array datasets and do not focus on the investigation of different methods and ratios of resampling before feature selection.

The researches presented in this section suggest the lack of studies, as well as the creation of mechanisms that are suitable for feature selection in fraud detection. Therefore, we investigated different resampling methods before feature selection step in order to develop more effective methods to identify frauds in e-commerce, which is the main contribution of this work.

3   Conceptual Fundamentals

In this section we present the conceptual foundations related with this work. In this research we refer to minority class (fraud) as positive and majority as negative.

In this work we report three supervised feature selection techniques, which are briefly described, as follows:

- **CFS** (Correlation-based Feature Subset Selection) aims to find sets of attributes that are highly correlated with the class, and they are not correlated with the other attributes in the set. Several metrics can be used to calculate the correlation matrix, in general the symmetric uncertainty is used [17].
- **Gain Ratio** is a method based on the concept of entropy of a sample space. The entropy of this space is characterized by the impurity of the data. Thus, the entropy returns a value from 0 to 1, depending on homogeneity of each attribute for classification [18].
- **Relief** considers that good attributes have equal values for instances of the same class and different values for instances of different classes. Relief as Gain Ratio perform the individual merit analysis for each attribute. It starts by randomly choosing an instance and find the closest instance of a different class (nearest miss) and nearest instance of the same class (nearest hit). After this, it defines the weight of each attribute as P(different values of A to an instance and their nearest miss) - P(different values of A to an instance and their nearest hit) [19].

The Gain Ratio and Relief feature selection techniques generate a ranking with merit of the attributes. In this work, we decide to not fix the number of attributes and we cut in the ranking through a threshold. This threshold was determined finding the elbow in the curve formed by a feature ranking.

To reduce the class imbalance to apply this feature selection methods we used resampling methods. There are two strategies to perform this task, *undersampling* and *oversampling*. In undersampling (US) the instances from the majority class are removed. The possible problem of this method is that important information can be lost. On the other hand, *oversampling* (OS) strategy duplicates instances of the minority class. This method can cause overfitting on the data.

The random methods, random undersampling (**RUS**) and random oversampling (**ROS**) do not care about the weaknesses of the strategies and randomly choose the instances that will be replicated or removed to deal with the imbalance between classes. However, there are smart methods that address these weaknesses. In this work, besides *ROS* and *RUS*, we used four smart resampling methods and create another one, specifically to apply before feature selection step to anomaly detection. These methods are briefly described in the Table 1.

Algorithm 1 shows the resampling method that we created specifically to reduce the class imbalance on feature selection step in anomaly scenarios. The main idea of this method is to remove the rare instances of negative class and replicate the rare instances of positive class, using the Smote method. Unlike

Table 1. Resampling methods

Method	Label	Type	Description	Reference
NearMiss-1	**NM-1**	US	Remove negative examples whose average distances to three closest positive examples are the smallest	[20]
NearMiss-2	**NM-2**	US	Remove negative examples based on their average distances to three farthest positive examples	[20]
NearMiss-3	**NM-3**	US	Remove negative examples to guarantee every positive example is surrounded by some negative examples	[20]
Smote	**Smote**	OS	Adds new artificial minority examples by interpolating between pre-existing minority instances	[21]
Sampling Outlier	**SO**	Mixed	Mixed undersampling and oversampling, following the Algorithm 1	Created by us in this work

the other methods, in this method it is not necessary to supply the new ratio between the classes as a parameter. In SO method, this ratio is determined by the number of instances that satisfy the conditions.

To compare the efficiency of the feature selection techniques and identify the frauds, we use three state of art supervised classification techniques, which are:

- **Bayesian Networks** are directed acyclic graphs representing dependency between variables of a probabilistic model, where the nodes are the attributes and the arcs represent the influence of relationships between variables. In general, the Bayesian Network (BN) is unknown, therefore we build the BN graph from the data. From the BN graph, we can determine the set of dependent variables and using the Bayes theorem to generate a conditional probability table. After this, we can find the probability of a event from the conditional probability table [22].
- **Logistic Regression** is a statistical technique that produces from a set of explanatory variables, a model that allows predicting the values taken by a categorical dependent variable. Thus, using a regression model, it is possible to calculate the probability of an event through a link function [23]. There are two classes of this link function, log-linear and logit. In this work we use the binary logistic regression model, it is a special case of the a generalized linear model (GLM) with the logit function [24].
- **Decision Tree - J48 Implementation** generates a classifier by a tree, this consists of leaf that indicates attributes, decision nodes that specifies a test

---

Algorithm 1. Sampling Outlier Resampling Method

```
 1  Give: i_pos and i_neg instances of positive and      16  M_F_pos = Mean(F_pos)
    negative class                                        17  DP_F_pos = Standard deviation of (F_pos)
 2  F_pos vector of size len(i_pos)                       18  M_F_neg = Mean(F_neg)
 3  F_neg vector of size len(i_neg)                       19  Dp_F_neg = Standard deviation of (F_neg)
 4  for i ∈ i_pos do                                      20  Cut_pos = (M_F_pos − Dp_F_pos)
 5  │   knn[i][]= Calculate the t KNN of i                21  Cut_neg = (M_F_neg − Dp_F_pos)
 6  │   for k ∈ Knn[i][] do                               22  for i ∈ i_pos do
 7  │   │   F_pos[k]+=1                                    23  │   if F_pos[i] < Cut_pos then
 8  │   end                                               24  │   │   Replicate i using SMOTE
 9  end                                                   25  │   end
10  for j ∈ i_neg do                                      26  end
11  │   knn[j][]= Calculate the z KNN of j                27  for j ∈ i_neg do
12  │   for k ∈ Knn[j][] do                               28  │   if F_neg[j] < Cut_neg then
13  │   │   F_neg[k]+=1                                    29  │   │   Remove j
14  │   end                                               30  │   end
15  end                                                   31  end
```

---

on the value of an attribute and a branch for each possible answer, which will lead to a sub- tree or a new leaf [25]. In this work we use the Decision Tree J48, it a builds decision trees from a set of training data using the concept of information entropy [26].

To evaluate the results achivied by classification techniques we use tree metrics, which are briefly described:

1. **AUC** is the area under the curve formed by false positive (FPR) rate and true positive rate (TPR), varying the classification threshold.
2. **Avg_F1** is a weighted average of the precision and recall, where an F-Measure reaches its best value at 1 and worst score at 0. This metric can be observed in Eq. 1. For each class (0 and 1) we get a F1 score, in other words the Eq. 1 returns one F1 for class 0 $(F1_{Class0})$ and other F1 for class 1 $(F1_{Class1})$, but we can get an Average F1 (Avg_F1) using Eq. 2.

$$F1 = 2 * \frac{precision * recall}{precision + recall} \tag{1}$$

$$Avg\_F1 = \frac{F1_{Class0} + F1_{Class1}}{2} \tag{2}$$

3. **EE** evaluate the financial gains after application of the fraud detection models, using the Eq. 3.

$$EE = k \cdot TN_{Value} - ((1 - k) \cdot FN_{Value} + p \cdot FP_{Value}), \tag{3}$$

where k is a gain of the company for each transaction; p is the penalty for false positive transaction; TN_{Value}, FN_{Value} and FP_{Value} are the sum of the transaction values for true negative, false negative and false positive transactions.

For a comparison of the models we describe the concept of maximum economic efficiency (EE_{Max}), which determines the highest amount that could be earned by the company in a perfect scenario, where an ideal model would get 100% precision and 100% recall for all transactions. That is, the model would predict success all fraud situations, and not identify any valid transaction as fraud. We also define the concept of Real Economic Efficiency ($E.E_{Real}$), which is the economic efficiency obtained by the company in this dataset.

However, we can not show results in monetary values due to the confidentiality agreement and privacy issues. For this reason, the results are presented in accordance with Relative Economic Efficiency ($Relative\_EE$), which measures the percentage gain on the current scenario of the company. Equation 4 defines the $Relative\_EE$ metric.

$$Relative\_EE = \frac{(EE - EE_{Real})}{(EE_{Max} - EE_{Real})} \qquad (4)$$

The term Economic Efficiency (EE) was used because in Web transactions involving financial values the costs of false positives and false negatives are not the same. It is estimated that the cost is approximately 3: 100. In other words, detecting a fraud is much more relevant than confirming a valid transaction, since the financial losses are much more significant for the first case [27].

4 Methodology

In this section we present the methodology used in this research to detect fraud in a Web transaction scenario. Figure 1 presents the process to investigate the research hypothesis. The main difference of this work is in steps 2 and 3, which are investigated and applied here.

The first step performs the **normalization of the database** to make easier the comparison and analysis of the techniques used herein. The continuous attributes were normalized between 0 and 1 and the discrete attributes were all considered as non-ordinal categories. After this, we separate a subset of training data, called **data for selection**, which will be used to select attributes for all feature selection techniques. This subset contains the same features, but different instances from the data that used for cross-validation, guaranteeing the generality of selection.

In the step 2 we started the validation of our first hypothesis, the high imbalance between classes reduces the effectiveness of selection attributes to detect fraud. To perform this **we used, before feature selection step, two distinct strategies about the proportion between the classes**. The first one generate a subset for feature selection with the real proportion between the classes, while in the second strategy we generate a subset for feature selection reducing the class imbalance through distinct resampling approaches.

To confirm this hypothesis would be enough to use just one resampling method, but one of the differences of this work is the comparison of resampling strategies before feature selection to identify frauds. Therefore, **we used 6**

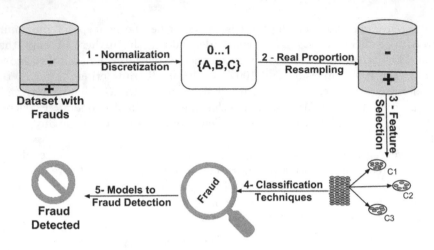

Fig. 1. Methodology to construct the fraud detection models

state of the art techniques and created a resampling technique, called *Sampling Outlier* (*SO*) with concepts that we considered important to feature selection in anomaly detection. These techniques were describes in Sect. 3.

The resampling methods, except *SO*, require the information about the new ratio between the classes. Thus, for each resampling method we create 8 subsets with distinct fraud proportion (5%,10%,15%,20%,25%,30%,40% and 50%). Therefore, we created 50 subsets to feature selection, formed through: 6 resampling methods with 8 distinct proportion of frauds, 1 resampling method that does not need to set the fraud ratio and 1 subset with the real proportion of fraud. Using these settings, we can determine the best fraud proportion for each technique and verify if increasing the fraud proportion the performance of the feature selection also increases or not.

In the step 3 **we applied the feature selection techniques**, presented in the Sect. 3, on each subset for feature selection. Besides that, we create a set of features with all features available (*NO_FS*) and 1 subset of attributes selected through the combination of the best attributes, selected in each feature selection techniques (*Merge*). In *Merge* the merit of a attribute is the frequency of this attribute on the subset of features with the best performances.

In the next step, **we used the classification techniques**, as described in Sect. 3, to identify fraud and evaluate the subsets of attributes for each approach. We use a training subset different from the feature selection training subset. To ensure the generality of the solutions we use the 8-fold Cross-Validation.

We do not use any resampling approach on training to classification, preserving the original fraud ratio. Although, we believe that the use of resampling also in this step could achieve better results, if we had used it we could be favoring any of the resampling methods applied on the feature selection step. To evaluate the results found in this step we used the metrics described in Sect. 3 and we performed the *Friedman* ([28]) and *Wilcoxon* ([29]) statistical test.

After the analysis, in step 5 **we constructed the fraud detection models**, which consists of a classification technique and a feature selection approach.

5   Case Study

This section presents our case study where we apply our models to detect fraud in a Web payment system.

5.1   Data Set Description

We used a real dataset from an electronic payment system called PagSeguro[1] to evaluate our methods. PagSeguro is one of the most used electronic payment systems in Latin America, mainly in Brazil. It is evident the need of efficient techniques that are able to identify frauds in this kind of system.

Table 2 shows the main information about this dataset.

Table 2. Dataset - General Overview

Number of features	381
Number of continuous features	248
Number of categorical features	133
Fraud proportion	$\cong 1.2\%$
Period of analysis	2012/2013

5.2   Experimental Results

In this subsection, we present and compare the results obtained after the use of the methodology explained in Sect. 4 applied in the real dataset from PagSeguro electronic payment system, from Subsect. 5.1.

Table 3 contains the number of features of each subset generated by the feature selection techniques, using the resampling or real proportion. After using the feature selection techniques, we obtain subsets of data with different numbers of features for each fraud proportion in all resampling methods, showing that the proportion of fraud directly influence the feature selection. The resampling method $NM - 3$, due to its nature has generated the same instances, regardless of the reported fraud ratio.

Table 4 summarizes the frequency of attributes on different quartiles, comparing the number of attributes present in subsets generated using resampling methods and subset generated from the real fraud ratio.

One important contribution of this work is the comparison between the different fraud ratio for each resampling method applied before feature selection

---

[1] http://pagseguro.uol.com.br.

Table 3. Feature Selection Approaches - Number of Features. Legend: □:CFS; △:GainRatio; ○:Relief

	NM-1			NM-2			NM-3			ROS			RUS			SMOTE			SO			Real		
	□	△	○	□	△	○	□	△	○	□	△	○	□	△	○	□	△	○	□	△	○	□	△	○
																			28	86	33	11	77	29
5%	34	325	26	46	192	31	27	44	32	22	81	26	20	74	30	28	67	26	-	-		-	-	
10%	19	334	19	43	191	28	27	44	32	24	77	31	27	52	23	27	69	32	-	-		-	-	
15%	20	340	26	32	21	22	27	44	32	29	45	32	32	52	37	22	81	31	-	-		-	-	
20%	19	340	23	28	23	25	27	44	32	28	57	31	38	63	30	23	78	34	-	-		-	-	
25%	18	341	26	28	26	17	27	44	32	27	56	30	28	70	29	22	84	31	-	-		-	-	
30%	19	341	26	26	31	27	27	44	32	27	34	31	25	42	25	21	80	38	-	-		-	-	
40%	18	45	20	22	41	20	26	37	30	23	37	27	23	51	26	21	37	27	-	-		-	-	
50%	14	56	17	22	44	18	26	41	22	26	35	27	28	12	22	20	54	28	-	-		-	-	

Table 4. Frequency of features selected by feature selection techniques, using a resampling method before feature selection or % the real fraud proportion.

Frequency	Number of features	
#	*Resampling*	**Real**
Present in any subset	375	107
Present in more than 25%	63	107
Present in more than 50%	12	9
Present in more than 75%	2	2
Present in all subset	0	2

step. Thus the graphic of Fig. 2 shows the AUC metric obtained after the classification on several subsets of attributes. This subsets were generated by each feature selection technique applied in the dataset with distinct fraud proportion in each resampling method and the real proportion (*Real*). Due to limited space we present this analysis only for the *AUC* metric, however we emphasize that we obtained similar behavior to analyze the $F1$ and EE metrics.

The graphs (*a*) in Fig. 2 show the behavior of resampling methods before the feature selection technique CFS. We can note that when we set an adequate fraud ratio, the *RUS* is a good alternative to the CFS technique. The *SO* method was better than 3 methods in all fraud ratios. All approaches that used resampling methods, before CFS feature selection, achieved better AUC than the real proportion (*Real*). Thus, we can prove that the high-class imbalance reduces the effectiveness of CFS feature selection technique and resampling methods before feature selection is a good strategy to improve the performance to identify frauds.

The graphs (*b*) show the same analysis for GainRatio. The Smote method was the best resampling for this technique. Contrary to CFS technique, the *SO* was not a good resampling method before GainRatio feature selection. When a good fraud ratio was used, all approaches using resampling methods, except *SO*, outperformed the approach using the *Real* before feature selection.

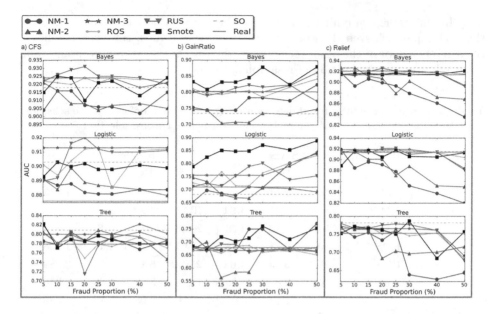

Fig. 2. Performance of classification techniques on subset of attributes selected by feature selection on dataset using resampling methods in distinct fraud ratio.

Lastly, the graphs (*c*) present analysis for Relief technique. We note that Relief was less sensitive to variation in fraud proportion. However, it selected more effective attributes when we used resampling before Relief Feature Selection. The exception were the approaches with resampling methods $NM - 1$ and $NM - 2$, which did not outperformed the *Real*. The method *SO* was very effective to reduce the class imbalance before feature selection with Relief.

In CFS and Relief techniques the subsets that used lower fraud ratios obtained better results. This behavior rejects the hypothesis that greater reduction in imbalance between classes implies better performance in the feature selection. We can note by Fig. 2 that the effectiveness of feature selection is not directly correlated with the increase of fraud ratio in resampling method.

In general, we can note for all feature selection techniques that the random undersampling (*RUS*) beats the random oversampling (*ROS*). This behavior agrees with [30], however in that work resampling is used before classification.

There is no one ideal fraud proportion for all resampling methods. We realize that each method has achieved better performance using a fraud ratio. Then, the method *SO*, created for us, can be a good alternative when the cost of performing a classification to choose the best fraud ratio is high. We choose the ideal fraud ratio for each resampling method, according to the graphic of Fig. 2, to reduce class imbalance before each feature selection technique.

Table 5 presents this fraud ratio for each resampling method and the percentage gain obtained in comparison with the same combination of feature selection and classification techniques, but using the real proportion before feature selection. In this table, we omit the results to the decision tree technique, because this techniques achieved the worst results to identify frauds. However, it follows the same behavior of the other techniques, when we analyze the percentage gain achieved after the application of resampling methods.

Table 5. Percentage gain in fraud detection using resampling before feature selection over the same techniques using the real proportion before feature selection.

FS	Resampling	%	Bayes			Logistic		
			AUC	F1	EE	AUC	F1	EE
CFS	NM-1	10	2.11	2.08	2.80	1.26	2.86	16.43
	NM-2	5	2.45	**3.61**	10.11	4.22	2.57	21.90
	NM-3	5:30	2.78	2.78	20.43	4.68	**4.86**	**59.08**
	ROS	20	2.78	2.78	10.97	4.11	3.43	12.68
	RUS	20	**3.56**	**3.61**	**21.29**	**5.02**	4.57	36.89
	Smote	10	2.89	3.06	11.18	3.08	2.86	3.46
	SO	X	2.11	2.08	2.80	3.08	4.14	13.26
Gain Ratio	NM-1	50	2.74	−5.86	23.02	18.14	9.84	**77.43**
	NM-2	5	−6.97	−1.85	3.77	5.06	0.00	25.66
	NM-3	50	7.47	**4.47**	33.96	18.85	5.24	31.42
	ROS	40	4.98	0.77	4.91	19.27	7.14	59.73
	RUS	25	1.74	−0.77	8.68	12.80	6.83	18.58
	Smote	30	**9.71**	−5.39	**45.66**	**25.04**	**12.70**	68.14
	SO	X	−8.47	0.00	36.98	−3.94	−0.63	26.99
Relief	NM-1	5	0.44	1.34	8.82	0.55	0.41	**5.23**
	NM-2	5	−5.03	−7.51	−28.36	−0.11	−4.38	−12.85
	NM-3	5:30	0.77	**3.49**	**19.96**	0.55	0.41	−2.18
	ROS	20	0.22	1.61	0.84	0.33	**0.68**	0.00
	RUS	20	1.31	2.82	17.65	**0.77**	0.27	4.36
	Smote	25	0.33	1.74	6.93	0.33	0.55	0.00
	SO	X	**1.42**	2.41	16.39	0.11	0.00	−8.28

We have highlighted in **bold** the best results for each classification technique in Table 5. We may note that using resampling for selecting attributes we achieved significant gains in terms of AUC, $F1$ and EE. The biggest gains were achieved when we used EE. That happens, because the cost of a True Positive (TP) and False Positive (FP) on this metric is not the same. Simulating what happens in real scenarios, the cost is 97% (TP) to 3% (FP). While in the other metrics used in this work the cost of a TP and FP is the same.

After the analysis of ideal fraud ratio for each resampling methods we created a feature selection approach that combines the features more frequent in the best subset selected from dataset with resampling methods, presented in the Table 5. This model, was called *Merge*. In addition, we create a approach that use all features to classification, that is, without the use of any feature selection in the dataset, this approach was called *NO_FS*.

Thus, we have 4 types of model to identify frauds. The models consist of the same classification techniques and distinct approach to feature selection or no feature selection. The model *Real* that do not use resampling before feature selection, the model *Resampling* that uses the best resampling method with the ideal fraud ratio for each feature selection technique, the model *Merge* that combines the features most frequent in the best subset of features and the model *No_FS* that contains all features available on dataset.

Table 6 presents the comparison of the best fraud detection models for each classification technique, using the three evaluation metrics. The models *Real* were excluded from this comparison, because in the previous analysis we demonstrated that when we have used resampling before feature selection we achieved better results. Thus, it compares the models using resampling on feature selection step, without feature selection and with the *Merge* feature selection strategy.

Table 6. Performance of models to fraud detection

	Bayes				Logistic				Tree			
		AUC	F1	EE		AUC	F1	EE		AUC	F1	EE
CFS	RUS	**0.931**	0.746	**0.564**	RUS	0.92	0.732	0.475	SO	**0.903**	**0.729**	0.393
GainRatio	Smote	0.881	0.614	0.386	Smote	0.889	0.71	0.38	Smote	0.753	0.689	0.395
Relief	RUS	0.926	**0.767**	0.56	RUS	0.917	0.737	**0.575**	RUS	0.776	0.713	**0.45**
Merge	-	0.918	0.743	0.539	-	**0.926**	**0.745**	0.435	-	0.774	0.689	0.39
No_FS	-	0.849	0.554	0.444	-	0.749	0.672	0.381	-	0.753	0.708	0.053

We can note in Table 6 that the Bayesian Network and Logistic Regression are more adequate techniques to detect fraud in this scenario than Decision Tree. The Feature Selection techniques Relief and CFS were more adequate to feature selection for fraud detection. The resampling method Random Undersampling, when correctly calibrated the fraud ratio, is a good resampling method to reduce the class imbalance on feature selection.

From Table 6 it is possible to infer that all models that use feature selection have achieved significant gains over the model *No_FS*. Thus, this analysis confirms the importance of feature selection in fraud detection. The method *Merge* can be a good strategy to select attributes to Logistic regression classification.

The model that combines Relief, using the random undersasmpling, and Logistic Regression, as classification technique, achieved the best economic gains. This model achieved economic gains of 57.5% in comparison to the real scenario of the Corporation, using just 8% of available features. In other words, if this model were used for fraud detection it would save 57.5% of fraud financial losses.

6   Conclusion

In this paper we analyze the use of feature selection to fraud detection in Web transactions. To perform this analysis we used three feature selection techniques and evaluate how the imbalance between classes can difficult this task. We used 7 resampling strategies in this step, including a resampling method created by us.

To identify frauds and measure the effectiveness of the approaches we build some fraud detection models that comprise a feature selection and a classification technique. In order to validate this methodology we perform experiments on a real database to detect fraud in a electronic payment system. To evaluate the models we use three performance metrics related with classification accuracy (F1 and AUC) and Economic Efficiency (EE).

Through the results of this research, we can validate our hypothesis that imbalance between classes reduces the effectiveness of the feature selection techniques to fraud detection. As a possible solution to this problem we use a distinct resampling strategy in the feature selection step. Our fraud detection model can improve by 57.5% the financial results obtained by the corporation.

In adition, in this work we show some interesting behaviors about feature selection to anomaly detection and resampling on feature selection step. The main conclusions are:

- The feature selection technique Gain Ratio was more sensitive to the imbalance between classes, while the Relief technique proved less sensitive to imbalance.
- Increasing the fraud ratio in resampling methods does not imply in increasing linearly the effectiveness of feature selection.
- Each resampling method is best suited to a distinct fraud ratio.
- Our method Sampling Outlier is a good alternative when the cost of performing a classification to choose the best fraud ratio is high.

Therefore, this research shows the need to provide mechanisms to increase the effectiveness of feature selection in anomalous scenario. Although we achieved good results with the used approaches, we believe that it is possible to further improve efficiency in fraud detection with some extensions of this research. Thus, we identify as a potential future work the proposal of new feature selection approaches, combined with resampling methods.

Acknowledgment. This research was supported by the Brazilian National Institute of Science and Technology for the Web (CNPq grant numbers 573871/2008-6 and 477709/2012-5), MASWeb (grant FAPEMIG/PRONEX APQ-01400-14), EUBra-BIGSEA (H2020-EU.2.1.1 690116, Brazil/MCTI/RNP GA-000650/04), CAPES, CNPq, Fapemig and Universo OnLine Inc. (UOL).

References

1. Bhattacharyya, S., Jha, S., Tharakunnel, K., Westland, J.C.: Data mining for creditcard fraud: a comparative study. J. Decis. Support Syst. **50**(3), 602–613 (2011)

2. Kim, K., Choi, Y., Park, J.: Pricing fraud detection in online shopping malls using a finite mixture model. Electron. Commer. Res. Appl. **12**(3), 195–207 (2013)
3. Almendra, V.: Finding the needle: a risk-based ranking of product listings at online auction sites for non-delivery fraud prediction. Expert Syst. Appl. **40**(12), 4805–4811 (2013)
4. Richhariya, P., Singh, P.K.: Article: a survey on financial fraud detection methodologies. Intl. J. Comput. Appl. **45**(22), 15–22 (2012)
5. Ravisankar, P., Ravi, V., Rao, G.R., Bose, I.: Detection of financial statement fraud and feature selection using data mining techniques. Decis. Support Syst. **50**(2), 491–500 (2011)
6. Kamal, A.H.M., Zhu, X., Pandya, A., Hsu, S., Narayanan, R.: Feature selection for datasets with imbalanced class distributions. Int. J. Softw. Eng. Knowl. Eng. **20**(02), 113–137 (2010)
7. Zhang, Y., Bian, J., Zhu, W.: Trust fraud: A crucial challenge for china e-commerce market. Electron. Commer. Res. Appl. **12**(5), 299–308 (2013)
8. Chiu, C., Ku, Y., Lie, T., Chen, Y.: Internet auction fraud detection using social network analysis and classification tree approaches. Intl. J. Electronic Commerce **15**(3), 123–147 (2011)
9. Keele, S.: Guidelines for performing systematic literature reviews in software engineering. Technical report, Ver. 2.3 EBSE Technical Report. EBSE (2007)
10. Chen, X., Wasikowski, M.: Fast: a roc-based feature selection metric for small samples and imbalanced data classification problems. In: Proceedings of the 14th ACM SIGKDD Conference on Knowledge discovery and data mining, pp. 124–132. ACM (2008)
11. Van Hulse, J., Khoshgoftaar, T.M., Napolitano, A., Wald, R.: Threshold-based feature selection techniques for high-dimensional bioinformatics data. Netw. Modeling Anal. Health Inform. Bioinform. **1**(1–2), 47–61 (2012)
12. Cuaya, G., Muñoz-Meléndez, A., Morales, E.F.: A minority class feature selection method. In: San Martin, C., Kim, S.-W. (eds.) CIARP 2011. LNCS, vol. 7042, pp. 417–424. Springer, Heidelberg (2011). doi:10.1007/978-3-642-25085-9_49
13. Alibeigi, M., Hashemi, S., Hamzeh, A.: DBFS: an effective density based feature selection scheme for small sample size and high dimensional imbalanced data sets. Data Knowl. Eng. **81**, 67–103 (2012)
14. Chawla, N.V.: Data mining for imbalanced datasets: An overview. In: Data Mining and Knowledge Discovery Handbook, pp. 853–867. Springer, Heidelberg (2005)
15. Van Hulse, J., Khoshgoftaar, T.M., Napolitano, A., Wald, R.: Feature selection with high-dimensional imbalanced data. In: IEEE International Conference on Data Mining Workshops, 2009, ICDMW 2009, pp. 507–514. IEEE (2009)
16. Maldonado, S., Weber, R., Famili, F.: Feature selection for high-dimensional class-imbalanced data sets using SVM. Inf. Sci. **286**, 228–246 (2014)
17. Hall, M.A.: Correlation-based feature selection for discrete and numeric class machine learning. In: Proceedings of the 17th International Conference on Machine Learning, ICML 2000, pp. 359–366. Morgan Kaufmann Publishers Inc., San Francisco (2000)
18. Kelleher, J., Namee, B.M.: Information based learning (2011)
19. Liu, H., Motoda, H. (eds.): Computational Methods of Feature Selection. Chapman and Hall, Boca Raton (2008)
20. Mani, I., Zhang, I.: kNN approach to unbalanced data distributions: a case study involving information extraction. In: Proceedings of Workshop on Learning from Imbalanced Datasets (2003)

21. Chawla, N.V., Bowyer, K.W., Hall, L.O., Kegelmeyer, W.P.: Smote: synthetic minority over-sampling technique. J. Artif. Intell. **16**, 321–357 (2002)
22. Maes, S., Tuyls, k., Vanschoenwinkel, B., Manderick, B.: Credit card fraud detection using bayesian and neural networks. Vrije Universiteir Brussel (2001)
23. Hosmer, D.W.: Applied Logistic Regression, 2nd edn. Wiley, New York (2000)
24. Dobson, A.J.: An Introduction to Generalized Linear Models. Chapman and Hall, London (1990)
25. Salzberg, S.: C4.5: Programs for machine learning by J. Ross Quinlan. Morgan Kaufmann Publishers, Inc., 1993. Mach. Learn. **16**(3), 235–240 (1994)
26. Quinlan, J.R.: C4.5: Programs for Machine Learning. Morgan Kaufmann Publishers Inc., San Francisco (1993)
27. Lima, R.A.F., Pereira, A.C.M.: Fraud detection in web transactions. In: WebMedia, pp. 273–280 (2012)
28. Friedman, M.: A comparison of alternative tests of significance for the problem of m rankings. Ann. Math. Stat. **11**(1), 86–92 (1940)
29. Benjamini, Y., Hochberg, Y.: Controlling the false discovery rate: a practical and powerful approach to multiple testing. J. R. Statistical Soc. Ser. B (Methodological) **57**, 289–300 (1995)
30. Drummond, C., Holte, R.C., et al.: C4. 5, class imbalance, and cost sensitivity: why under-sampling beats over-sampling. In: Workshop on Learning from Imbalanced Datasets II, vol. 11. Citeseer (2003)

Business Processes, Web Services and Cloud Computing

Multimodal Indexing and Search of Business Processes Based on Cumulative and Continuous N-Grams

Hugo Ordoñez[1]([✉]), Armando Ordoñez[2], Carlos Cobos[3], and Luis Merchan[1]

[1] Research Laboratory in Development of Software Engineering,
Universidad San Buenaventura, Cali, Colombia
{haordonez,lmerchan}@usbcali.edu.co

[2] Intelligent Management Systems, Fundación Universitaria de Popayán,
Popayán, Colombia
armando.ordonez@docente.fup.edu.co

[3] Information Technology Research Group (GTI),
Universidad del Cauca, Popayán, Colombia
ccobos@unicauca.edu.co

Abstract. Reuse of business processes may contribute to the efficient deployment of new services. However, due to the large volume of process repositories, finding a particular process may become a difficult task. Most of the existing works in processes search are focused on textual information and graph matching. This paper presents a multimodal indexing and search model of business processes based on cumulative and continuous n–grams. The present method considers linguistic and behavior information represented as codebooks. Codebooks describe structural components based on the n-gram concept. Obtained results outperform the precision, recall and F-Measure of previous approaches considerably.

Keywords: Business process models · Multimodal search · Cumulative and continuous n-grams · Repository · Evaluation

1 Introduction

Reuse of components such as Web services and business processes may significantly contribute to the quick and efficient deployment of new services. The latter is crucial to the creation of value-added services, leading to competitive differentiation and customer retention. To achieve this goal, organization's tasks should be grouped around business functions such as marketing, sales, production, finance, customer service. Each one of these business functions may be carried out separately according to its business model [1].

Commonly, company's organizational guidelines are modeled using Business Processes (BP) formed by procedures or activities that aim at achieving a common objective [2]. BP models may be used to document and share organization's internal procedures; guide the development of new products and support improvement processes, among other functions. These business processes

© Springer International Publishing AG 2017
D. Bridge and H. Stuckenschmidt (Eds.): EC-Web 2016, LNBIP 278, pp. 129–141, 2017.
DOI: 10.1007/978-3-319-53676-7_10

are commonly stored in repositories containing hundreds or even thousands of process models, which in turn are made up of tens or hundreds of elements (tasks, roles and so on). Due to the large volume of BP models, BP management of repositories may become a great challenge. Furthermore, BP models are generally used by many stakeholders from diverse areas, with different skills, responsibilities, and objectives. Additionally, sometimes these BP are distributed in different and independent organizational units [3].

Consequently, finding a particular BP that matches specific requirements may be a complex task that requires considerable time and effort. To address this issue, many researchers have developed different mechanisms to search and retrieve business processes [4]. Most of these mechanisms start with user queries expressed as a complete business process or a fraction thereof, and then a set of similar processes is extracted from the repository, commonly these retrieved processes can be used as a starting point for creating new models according to new user's needs.

Most of the existing approaches in this field are based on typical measures such as linguistics, structure, and behavior. However, other techniques from the field of Information Retrieval (IR) may be applied to improve existing results. Among these IR techniques, the multimodal search reports very good results among users, this is in part because multimodal search combines different information types to increase the accuracy of search [5].

On the other hand, the use of n-grams in information retrieval systems for indexing and searching provides the following advantages [6]. (i) simplicity: due to the fact that linguistic knowledge and resources are not needed. (ii) robustness: since this approach doesn't depend highly on orthography variations and grammatical errors. (iii) independence of the domain and topic of the information to retrieve. (iv), efficiency, as the information processing is done in a single step. Due to the reasons above, N-grams offer advantages to the information retrieval approaches based on characters or complete words.

The approach in this paper presents a multimodal indexing and search model of business processes based on cumulative and continuous n–grams; the model unifies linguistic and behavioral information features in one search space.

The present approach improves the results obtained using the mechanism described in [7]. In order to achieve this, this method considers two dimensions of processes information: firstly, linguistic information which includes names and descriptions of BP elements (e.g. activities, interfaces, messages, gates, and events), and secondly, behavior information represented as codebooks in the form of text strings. Codebooks include all the structural components representing sequences of the control flow (i.e. the union of two or more components of the control flow simulates the formation of cumulative continuous N-grams).

The validation process of the approach was done using a BP repository created collaboratively by 59 experts; this repository is described in [8]. Equally, the approach was compared with other algorithms. In order to evaluate the relevance of the results, some measures from information retrieval domain were used.

The rest of this paper is organized as follows: Sect. 2 presents the related work, Sect. 3 describes the proposed multimodal search and indexing model, Sect. 4 presents the evaluation, and finally Sect. 5 describes conclusions and future works.

2   Related Works

Most of the existing approaches in this field use information retrieval techniques based on linguistic [9], association rules [10] and genetic algorithms [11]. In order to perform the search, these approaches use a set of elements or data types existing in business processes.

For example, linguistic-based approaches use textual information during the search process. This textual information may be the name or description of activities, events, and logic gates. Some techniques in this field use Vector Space Model (VSM) representations with a frequency of terms (TF) and cosine distance [9] to generate rankings of relevant results.

Approaches based on association rules analyze previous executions of business processes stored in log files. During the search, some phrases related to business process activities are identified, to do so, domain ontologies are used; additionally, some activity patterns are also identified. To create a list of results, a heuristic component that determines the frequency of detected patterns is employed [10].

For its part, approaches based on genetic algorithms use formal representations of the processes (graphs or state machines) and include additional data such as the number of inputs and outputs per node, edge labels, nodes name or description. Although this method may achieve precise results, execution time may be very high [11].

Finally, some approaches use repositories of business processes described and stored using XML. In order to perform the search process, these methods use languages like IPM Process Query Language (PQL - IPM). This language supports specific BP queries that make it possible, for example, to retrieve processes containing specific activities, transitions or connection between activities [12].

Most of the existing works in this area are limited to match inputs and/or outputs, using textual information of business processes elements. However, these approaches ignore relevant information of business processes such as execution flow, behavior, structure, activity type, gate type and event type [13,14].

The present paper describes an approach for multimodal indexing and search of BP. The approach integrates textual information of BP components (e.g. names and description of tasks, events, and gates) and codebooks. The codebooks are formed by joining information of BP sequential behavior, (e.g. task - task, task - gate, task - event - gate, among other types of sequential behavior). This type of search allows having a better search space and generating more accurate results (these results are organized according to the similarity to the query).

3   Multimodal Indexing and Search of Business Process

The main tasks of the approach are Indexing (a new BP is going to be added to the repository) and search (user want to seek a BP similar to a specific query). Next, both tasks are described.

3.1   Indexing Task

This task is executed before the BP models are indexed and stored in the repository. This task includes textual processing and the generation of a search index. Next, the modules responsible for implementing these rules are described (See Fig. 1).

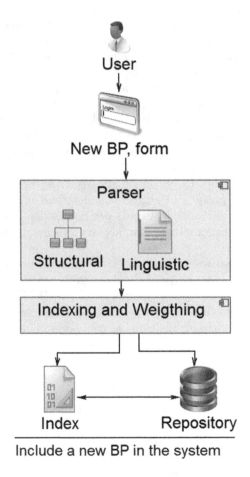

Fig. 1. Modules of the Indexing task

3.1.1   New BP Form

This component is used by the graphical user interface (GUI) and provides interaction with the user (to add, modify and delete BP models from the repository and the index). BPs in the repository are stored in their original XML format using BPMN (Business Process Modeling Notation) [15].

3.1.2   Parser

The parser implements an algorithm that takes a BP described in BPMN notation and builds linguistic and structural components (codebooks); this component also generates a search index consisting of two arrays: an array MC of textual features and a matrix MCd of structural components. The algorithm is described below.

Formation of linguistic component (Linguistic): suppose that we have a repository T of BPs: $T = \{BP_1, BP_2, ..., BP_i, ..., BP_I\}$ where I is the total number of BPs that it contains. The first step of the algorithm is to read each BP_i and to represent it as a tree A_i. Then the algorithm takes each A_i, extracts the textual characteristics (activity name, activity type, and description) and forms a vector $Vtc_i = \{tc_{i,1}, tc_{i,2}, ..., tc_{i,l}, ..., tc_{i,L}\}$, where L is the number of textual characteristics $tc_{i,l}$ found in A_i. For each vector Vtc_i, which represents a BP_i, a row of a matrix of textual components MC is constructed. This row contains the linguistic (textual) component for the BP stored in the repository. In this matrix, i represents each BP and l a textual characteristic for each of them.

Formation of codebook component (Structural): a codebook Cd is a set of N structural components describing one or more BP nodes in the form of text strings. The set of codebooks formed from the whole repository is called the codebook component matrix. This matrix is formed by taking each tree A_i, which all contain a vector of codebooks $Vcb_i = \{Cd_{i,1}, Cd_{i,2}, Cd_{i,k}....Cd_{i,p}\}$. Therefore, the codebook component matrix $MCd_{i,k}$ is formed where i represents the current BP and k represents its correspondent codebook.

For example, Fig. 2 shows a fragment of a BPi with its activities. Each activity is represented by a text string defining the node type (StartEvent, TaskUser, and TaskService). The node type refers to the functionality of each activity within the BP.

Codebooks are formed simulating the technique of traditional n-grams. These codebooks are sequences of textual elements: words, lexical item, grammatical labels, etc. arranged according to the order of appearance in the analyzed information. This method differs from previous works where traditional n-grams are formed with two components (N = 2, bigrams) [7]. Conversely, in the present approach, the representation includes n-grams with N = 1 (unigrams), N = 2 (bigrams), N = 3 (trigrams) and so on until a maximum value N = M.

The present approach uses continuous and cumulative n-grams. Continuous and cumulative n-grams are text fragments which may have diverse sizes within the continuous text, and they may be composed of sequential elements within

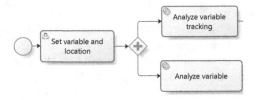

Fig. 2. Types of component in Business process

the text. N-grams may be convenient for the tree based representation of business processes. Next, a sample of BP is shown in Fig. 2, equally correspondence between BP components in Fig. 2 and their node types are presented in Table 1. This correspondence is done using text strings.

The codebook of n-grams representing the example process (described in Fig. 2) is composed of n-grams which vary in size from 1 to 4 (M = 4). n-grams of size 1 are: StartEvent, TaskUser, RouteParallel, TaskService, TaskService, on the other hand, n-grams of size 2 are formed as described in Fig. 3: $\{StartEvent\_TaskUser_1,\ TaskUser\_RouteParallel_2,\ RouteParallel\_TaskSer - vice_3,\ RouteParallel\_TaskService_4\}$, where $StartEvent\_TaskUser_1$ corresponds to the concatenation of Star Event with the Set variable and location activity, similarly to the other components.

Table 1. Correspondence between BP components in Fig. 2 and their types

Component	Type Representation
Start	Start Event
Set variable and location	TaskUser
Route	RouteParallel
Analyze variable tracking	TaskService
Analyze variable	TaskService

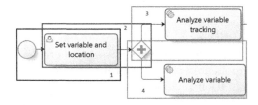

Fig. 3. Formation of codebooks of size 2

Codebooks of size 3 are shown in Fig. 4: $\{StartEvent\_TaskUser\_Route-Parallel_1,\ TaskUser\_RouteParallel\_TaskService_2,\ TaskUser\_RouteParallel\_TaskService_3\}$, as can be seen, as n-gram size grows a bigger part of the sequential path is covered (by concatenating its components).

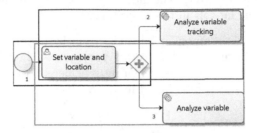

Fig. 4. Formation of codebooks of size 2

Codebooks of size 4 (see Fig. 5) are: $\{StartEvent\_TaskUser\_RouteParallel\_TaskService_1,\ StartEvent\_TaskUser\_RouteParallel\_TaskService_2\}$ Unlike traditional n-grams, these codebooks provide a better and higher representation of processes control flow and behavior semantics. This is because these codebooks integrate components of the sequences in the control flow. Behavior semantics of business processes describes the activities and its execution order [12]. It is important to note that as codebooks increases in size, they represent better the behavior semantics of processes.

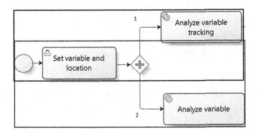

Fig. 5. Formation of size 4 codebook

Finally the codebooks vector $Vcd_i = $ {StartEvent, TaskUser,RouteParallel, TaskService, TaskService, StartEvent_TaskUser, TaskUser_RouteParallel, RouteParallel_TaskService, RouteParallel_TaskService, StartEvent_TaskUser_Route-Paralle, TaskUser_RouteParallel_TaskService,TaskUser_RouteParallel_TaskService, StartEvent_TaskUser_RouteParalle_TaskService, StartEvent_TaskUser_RouteParalle_TaskService}.

The cumulative and continuous n-grams concept can also be used for terms (linguistic features) presented in BPs. However, in the present work, these n-grams are used only for the behavioral features which form the component MCd.

3.1.3 Indexing and Weighting

Indexing: In this component, the linguistic and codebook components are joined and weighted to create a multimodal search index MI composed of the matrix of the linguistic component (MC) and the codebook component matrix (MCd) i.e. MI = {MCd U MC}. The index also stores a reference to the physical file of each one of the models in the repository.

Weighting: Next, this component built the term by document matrix applies a weighting scheme of terms similar to that proposed in the vector space model for document representation. This weighting scheme is based on Eq. 1, initially proposed by Salton, that can be found in the work of [16].

$$W_{i,j} = \frac{F_{i,j}}{Max(F_i)} * \log\left(\frac{N}{N_j + 1}\right) \tag{1}$$

In Eq. 1, $F_{i,j}$ is the observed frequency of a component j in BP_i, the component j can be a linguistic or codebook component. $max(F_i)$ is the greatest frequency observed in BP_i, N is the number of BPs in the collection, and N_j is the number of BPs in which the linguistic or codebook component j appears. Each cell $W_{i,j}$ of the multimodal index matrix reflects the weight of a specific element j of a BP_i compared to all the elements of the BP in the repository.

Index: The multimodal index is stored in a physical file within the file system of the operating system, in which each BP is indexed through a pre-processing mechanism. This mechanism converts all the terms of the linguistic and codebook component matrices to lowercase and removes stop words, accents, and special characters. Subsequently, the stemming technique (Porter algorithm [16]) is applied. This technique transforms the matrix elements into their lexical root (e.g. the terms "fishing" and "fished" become the root, "fish") and stores in the repository a reference to the physical file of each BP.

Repository: The repository is the central BP storage unit, similar to a database, that shares information about the artifacts produced or used by a company [17]. It is responsible for storing and representing all the attributes of information present in the BP (roles, description of activities, timers, messages, and service calls) [18]. Currently, the repository is composed of 100 BPs obtained from real processes of diverse companies of the telecommunications and georeferencing domains.

3.2 Search Task

This task is responsible for performing BP searches. These searches may have three query options: linguistic, codebook, and multimodal. Each query is represented through a terms vector $q = \{t_1, t_2, t_3, \ldots t_j \ldots t_J\}$. The same pre-processing mechanism applied in the indexing task (parser) is applied to the BP query, thus obtaining the terms of the query vector reduced to their lexical root (See Fig. 6).

Query forms: In this component, the user has forms corresponding to the graphical user interface (GUI). These forms are used to execute searches and display result lists.

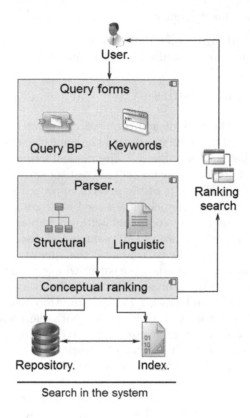

Search in the system

Fig. 6. Search task

Conceptual ratings: Once the vector is pre-processed (using the Parser process previously presented in Sect. 3.1.2), the search is executed according to the query option chosen by the user. Next, the conceptual rating (score) expressed by Eq. 2 (based on the Lucene practical scoring function) is applied. This rating is used to return a list of sorted results.

$$\text{Score} = (q, d) = \text{coord}(q, d) * \sum_{t \in q} (\text{tf}(t \in d) + \text{idf}(t)^2 * \text{norm}(t, d)) \qquad (2)$$

In Eq. 2 t is a term of query q; d is the current BP being evaluated with q; $\text{tf}(t \in d)$ is the term frequency, it is defined as the number of times the term t appears in d. BPs will be later ranked according to the values of term frequency scores. idf(t) is the number of BPs in collection divided by the number of BPs in which the term t occurs (inverse frequency). coord(q, d) is the scoring factor based on the number of query terms found in the queried BP (those BPs containing the most query terms are scored highest). norm(t,d) is the weighting factor in the indexing, taken from $W_{i,j}$ in the multimodal index.

List of results (Ranking): Once the results are sorted and filtered, they are listed according to the level of similarity to the query BP.

4   Evaluation and Results

Results obtained with the present approach were compared with the results of the manual evaluation. This comparison used the same elements (query BPs) and evaluation methodology presented in [8]. The manual evaluation was performed using a closed test set created by experts. This test set was created collaborative by 59 experts [8]. Also, the results of the present approach were compared with the results of MultiSearchBP model (it is based on 2-grams, n = 2, described in [7]). Finally, the results were compared with an implementation of the A* algorithm [19] (which is widely used in BP search and discovery).

In this sense, it is possible to evaluate the quality of results obtained from query execution using statistical measures widely used in the information retrieval field. The metrics/measures are: graded effectiveness: graded precision (Pg), graded Recall (Rg) and graded F-Measure. Moreover, the ranking generated using the present approach is compared with the ranking generated by human experts. To do this latter comparison A(Rq) [20] measure was used. A(Rq) was used to determine the degree of coincidence of the position of each BP in each one of the rankings generated by each request.

Figure 7 presents the results obtained in the evaluation process, regarding the graded precision, the proposed model (continuous N-grams) achieved 97.33%, exceeding by 3.33% the MultiSearchBP (2-grams) and by 25.33% the A* algorithm. The latter allows evidence that representation of BP flows using cumulative and continuous n-grams, allows more precise queries. The latter is because these are formed following the continuous path of the trees that represent the processes. Consequently, this approach leads to better results because the query BP and recovered BPs have similarity in fragments of higher size in their execution flow, but MultiSearchBP (2-grams) only compare fragments of two components in the control flow of the BP.

MultiSearchBP and A* obtained low values of graded Recall. A* obtained 26%, MultiSearchBP (2-grams) 30% and continuous N-grams 44%. The latter results because all methods generate rankings limited to a maximum of ten (10) results, leaving out other BP that may be relevant to the query. By its part, continuous N-grams exceeds MultiSearchBP (2-grams) by 14% and A* 11% for the whole set of queries.

The graded F-measure represents the harmony between the results of Pg and Rg. On average, continuing N-grams exceeds the performance of MultiSearchBP (2-grams) by 15% and A* by 22%. This result allows evidence that the rankings generated by the multimodal indexing and search model based on cumulative and continuous N-grams have a high level of similarity with the collaborative ranking generated manually by experts [8].

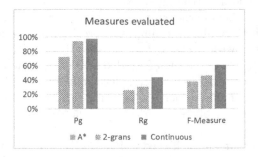

Fig. 7. Results in measures evaluated

Figure 8 shows the level of agreement A(Rq) between the ideal ranking generated by evaluators and the automatic ranking generated by our method. It can be observed that for each query the proposed approach (continuous and cumulative N-grams) generated classifications that match considerably with those generated by experts (ideal ranking). For example, in Query 2 (Q2), the proposed method reached 88%, while MultiSearchBP reached 83% and A* reached a maximum of 68% at its highest point.

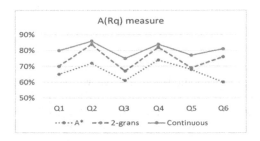

Fig. 8. Ranking concordance for each query (A(Rq))

Finally, regarding the ranking of overall similarity (considering all queries), the continuous and cumulative n-grams approach reached 81%, MultiSearchBP (2-grams) 74% and A* 66%. The latter indicates an increase in the quality of the generated ranking when the multimodal indexing and search based on cumulative and continuous n-grams is applied. The proposed method retrieves the most relevant BP for each query and avoids retrieving processes not relevant to the user's query.

5 Conclusions and Future Works

This paper presents an approach for improving the recovery of business processes (BP) based on multimodal indexing and search using cumulative and continuous n-grams [6] of behavioral (structural) features. The use of textual information and structural information in the multimodal index through cumulative and continuous n-grams offers greater flexibility and precision in queries. The proposed model was evaluated with a closed test set of BPs created collaboratively by experts. The model reaches levels of graded precision between 91% and 98%. Regarding performance of search process determined using F-measure, the proposed model exceeds in 25% other mechanisms (MultiSearchBP (2-grams) and A*), which demonstrates coherence in the relevance of results obtained with Pg and Rg.

Despite the improvements achieved in Pg and Rg using the present approach, the formation of the cumulative continuous n-grams increases the order of the algorithm. Consequently, the indexation and search time increases too. This situation arises because the algorithm must analyze the control flow of the BPs in the repository and the query several times. The latter process is required to form the n-grams.

Future works include the integrating of clustering algorithms (like K-means, Clicks, Start, and Fuzzy C–means) to the model and compare the groups created with other results reported in the state of the art. Equally, specific domain ontologies will be integrated to enrich both BP and queries semantically. On the other hand, it is planned to develop an automatic evaluation module that generates graphs and relevance measures. Finally, the evaluation will be expanded by applying new measures for the BP search.

References

1. Reijers, H.A., Mans, R.S., van der Toorn, R.A.: Improved model management with aggregated business process models. Data Knowl. Eng **68**(2), 221–243 (2009)
2. Gröner, G., Bošković, M., Silva Parreiras, F., Gašević, D.: Modeling and validation of business process families. Inf. Syst. **38**(5), 709–726 (2013)
3. La Rosa, M.: Detecting approximate clones in business process model repositories Business Process Model Repositories, April 2015
4. Dumas, M., Garcí-bañuelos, L., La Rosa, M.: BPMN Miner: automated discovery of BPMN process models with hierarchical structure, December 2015
5. Caicedo, J.C., BenAbdallah, J., González, F.A., Nasraoui, O.: Multimodal representation, indexing, automated annotation and retrieval of image collections via non-negative matrix factorization. Neurocomputing **76**(1), 50–60 (2012)
6. Vilares, J., Vilares, M., Alonso, M.A., Oakes, M.P.: On the feasibility of character n-grams pseudo-translation for Cross-Language Information Retrieval tasks. Comput. Speech Lang. **36**, 136–164 (2016)
7. Ordoñez, H., Corrales, J.C., Cobos, C.: Business processes retrieval based on multimodal search and lingo clustering algorithm. IEEE Lat. Am. Trans. **13**(9), 40–48 (2015)

8. Ordoñez, H., Corrales, J.C., Cobos, C., Wives, L.K., Thom, L.: Collaborative evaluation to build closed repositories on business process models, pp. 311–318 (2014)
9. Koschmider, A., Hornung, T., Oberweis, A.: Recommendation-based editor for business process modeling. Data Knowl. Eng. **70**(6), 483–503 (2011)
10. Hernandez-Gress, N., Rosso-Pelayo, D.A., Trejo-Ramirez, R.A., Gonzalez-Mendoza, M.: Business process mining and rules detection for unstructured information. In: Ninth Mexican International Conference on Artificial Intelligence (2010)
11. Mehnen, J., Turner, C.J., Tiwari, A.: A Genetic Programming Approach to Business Process Mining (2010)
12. Yan, Z., Dijkman, R., Grefen, P.: Business process model repositories – Framework and survey. Inf. Softw. Technol. **54**, 380–395 (2012)
13. Rivas, D.F., Corchuelo, D.S., Figueroa, C., Corrales, J.C., Giugno, R.: Business process model retrieval based on graph indexing method. In: Muehlen, M., Su, J. (eds.) BPM 2010. LNBIP, vol. 66, pp. 238–250. Springer, Heidelberg (2011). doi:10.1007/978-3-642-20511-8_24
14. Grigori, D., Corrales, J.C., Bouzeghoub, M., Gater, A.: Ranking BPEL processes for service discovery. IEEE Trans. Serv. Comput. **3**(3), 178–192 (2010)
15. Chinosi, M., Trombetta, A.: Computer standards & interfaces BPMN: an introduction to the standard. Comput. Stand. Interfaces **34**(1), 124–134 (2012)
16. Manning, C.D., Prabhakar, R., Schütze, H.: An Introduction to Retrieval Information, 2nd edn. (2008)
17. Dijkman, R., Gfeller, B., Küster, J., Völzer, H.: Identifying refactoring opportunities in process model repositories. Inf. Softw. Technol. **53**(9), 937–948 (2011)
18. Reimerink, A., de Quesada, M.G., Montero-Martínez, S.: Contextual information in terminological knowledge bases: a multimodal approach. J. Pragmat. **42**(7), 1928–1950 (2010)
19. Grigori, D., Corrales, J.C., Bouzeghoub, M.: Behavioral matchmaking for service retrieval: application to conversation protocols. Inf. Syst. **33**(7–8), 681–698 (2008)
20. Guentert, M., Kunze, M., Weske, M.: Evaluation Measures for Similarity Search Results in Process Model Repositories, pp. 214–227 (2012)

Scoring Cloud Services Through Digital Ecosystem Community Analysis

Jaume Ferrarons[1]([⊠]), Smrati Gupta[1], Victor Muntés-Mulero[1],
Josep-Lluis Larriba-Pey[2], and Peter Matthews[1]

[1] CA Strategic Research Labs, CA Technologies, Barcelona, Spain
jaume.ferrarons@ca.com
[2] Universitat Politécnica de Catalunya - BarcelonaTech, Barcelona, Spain

Abstract. Cloud service selection is a complex process that requires assessment of not only individual features of a cloud service but also its ability to interoperate with an ecosystem of cloud services. In this position paper, we address the problem by devising metrics to measure the impact of interoperability among the cloud services to guide the cloud service selection process. We introduce concrete definitions and metrics to contribute to measuring the level of interoperability between cloud services. We also demonstrate a methodology to evaluate the metrics via a use case example. Our contributions prove that the proposed metrics cover critical aspects related to interoperability in multi-cloud arena and therefore form a robust baseline to compare cloud services in systematic decision making environments.

Keywords: Cloud service · Interoperability · Multi-cloud · Scoring · Decision support

1 Introduction

The growth of e-business and application based economies has impacted multiple facets of IT industry and specially the cloud computing domain. E-businesses heavily rely on cloud architectures to provide light-weight, economic, easy to manage services in terms of infrastructure, platform, software and so on. Cloud market is growing and changing fast, it is forecasted to grow a 16.5% in 2016 [11]. In this cloud market, different cloud services offer different features to suit the requirements of the user. Nonetheless, the value of a particular service is not only based on the quality of its individual features and specifications, but also on the quality of the other services able to support this particular service and extend its individual set of features, either from the same cloud provider or a different one. For this purpose, the interoperability among cloud services has to be taken into account to make a cloud service selection because the selected set of cloud services should not only meet the requirements of the application independently, but also the requirements to interoperate among them to ensure a smooth execution of the application [5]. Because of the lack of standards and

© Springer International Publishing AG 2017
D. Bridge and H. Stuckenschmidt (Eds.): EC-Web 2016, LNBIP 278, pp. 142–153, 2017.
DOI: 10.1007/978-3-319-53676-7_11

the lack of universal interoperability imposed by companies concerned about uncontrolled customer churning, not all the services are supported by the same complementary services. At the same time, different services from different CSPs may be interoperable with the same third-party service. Understanding these aspects of a particular service may be essential when planning to migrate to an equivalent service. In particular, the stickiness of a specific service to the *wrong* services may impede the migration process to other services and may make it difficult to avoid vendor lock-in problems, jeopardizing flexibility and agility and even causing extra costs and lack of productivity. For example, let us suppose a cloud DB has been selected along with other cloud services, in order to sustain an application, ignoring their ability to interact with other services. In this case, the replacement of the cloud DB, for example due to a new policy on data location, becomes more challenging because other cloud services in use such as backup, load balancing or monitoring cloud services may not be compatible with another cloud DB. Therefore, not only the service affected has to be replaced but also all the services that are not compatible with the new cloud DB.

Scenarios in which several cloud services from different providers are used in conjunction offer multiple benefits like: reduced dependency on a single vendor and prevention of vendor lock-in, or freedom to leverage different features provided by the different vendors for different components. These scenarios also impose two important challenges: complexity and heterogeneity management. Complexity is a consequence of the vast growth in the number of options available when connecting a new service to an existing ecosystem and establishing contractual relations with the cloud service providers. Heterogeneity is caused by the usage of services from different cloud providers to sustain the application.

Therefore, in order to fully leverage the benefits of the multi-vendor cloud service landscape, we need to provide mechanisms to cope with complexity and heterogeneity. Cloud service selection decision making tools may help tackling these challenges. The evaluation of a cloud service should not only be performed on the basis of the features it offers but also on its capacity to interact with other services. Many research works in the topic of cloud service selection either select one cloud service to deploy an application [2,6,8,9] or deal with workload balancing and migration among cloud services [5,10]. The nature of the information used in each of these works is different. In [8], authors compare the results obtained when using different Multi-Criteria Decision-Making (MCDM) methods for selecting an Infrastructure as a Service (IaaS). The criterion they use is based on price, memory, CPU and I/O performance. Authors of [6] propose a combined methodology for cloud service selection based on objective performance data and subjective information from the customers to include their vision on the Quality of Service obtained. The system presented in [9] is focused on recommending cloud services using information about memory, storage transfer times and financial costs. However, most of these works lack an important aspect: they do not address the joint selection of a set of cloud services such that the cloud service set can interoperate with each other agnostic of the vendors. It is also the case of many works done in the web service composition field [7], in which web service selection processes are based on: performance, reliability, capacity, protocols etc. disregarding the repleacebility and interoperability of this and other components.

Some authors [5] point out the risks of multi-cloud environments based on: the difficulty of replacing a service, the cloud service interoperability of the components in use and the security aspects like data confidentiality and integrity risks. On the other hand, a holistic Decision Support System (DSS) framework is presented in [2]. This framework is based on risk assessment, but also considering the specific characteristics of each cloud service, their cost and the quality of the solution. However, these works do not provide a concrete methodology to define and analyze the interoperability. To the best of our knowledge, these issues have been largely ignored in the literature.

In this paper we address the lack of definitions and methods to measure interoperability among the cloud services in the following ways: firstly, we propose a set of formal definitions to establish the basis for a framework to objectively assess the level of interoperability of a service with respect to other services in a digital ecosystem. Secondly, specific metrics useful to assess the interoperability level in multi-cloud environments based on three dimensions: cardinality, quality and functional diversity. Thirdly, we propose a concrete methodology to evaluate these metrics via a use case example thereby eliciting a method to embed such metrics in decision support tools for cloud service selection.

The remainder of this paper is organized as it follows: Sect. 2 contains the definitions and metrics used to assess the level of interoperability of cloud services. The experimental setup is exposed in Sect. 3 along with the numerical results obtained. Finally, Sect. 4 presents some conclusive remarks and outlines future lines of work.

2   Definitions and Metrics

In order to tackle the challenges that emerge in a multi-cloud environment, we evaluate the cloud services on the basis of their ability to interoperate with other services thereby assessing their capacity to exist in multi-cloud environment.

2.1   Definition of Interoperability

According to TOSCA standard [1] interoperability is defined as *"the capability for multiple components to interact using well-defined messages and protocols. This enables combining components from different vendors allowing seamless management of services."* The capability of a service to be interoperable with other services brings two major advantages: first, it allows the ability to utilize the features provided by the cloud service within an application agnostic of the presence of other cloud services. Second, it allows extending the features of a cloud service with the functionality provided by other services. Therefore, the more interoperable is a cloud service with other services, the larger the customer capacity to leverage its features and functionalities.

2.2   Metrics to Measure Cloud Service Interoperability

We now define the notation that will be further used to define the metrics to measure the cloud service interoperablity. Let us define CS as the pool of cloud services available in the market to be considered for cloud service selection in a particular use case. TOSCA establishes that cloud services can be grouped into different cloud service types based on the functionality they provide, for example: cloud databases, compute instances or logs analyzers. Let T denote the set of all the service types under consideration such that $T = \{a, b, ...\}$. Let us also define CS_t as the set of services in CS that are of type $t \in T$. The instances in set CS_t are denoted by x^t, y^t, etc. For example, CS_a is the set of cloud services of type a where a may refer to *cloud database* type. CS_a will contain all the instances of the DBaaS type of service such as: Cloud SQL from the Google Cloud Platform[TM], Amazon Relational Database Service from Amazon Web Services[TM], etc. The set of cloud services CS can be seen as the union of the disjoint sets of cloud services of different types $CS = CS_a \cup CS_b \cup ... \cup CS_m$.

Let us denote two cloud service instances of type a and b as x^a and y^b such that $a, b \in T$ and $x^a \in CS_a$ and $x^b \in CS_b$. We denote the interoperability between services x^a and y^b as $x^a \leftrightarrows y^b$. In case these two services are not interoperable, we denote it as $x^a \not\leftrightarrows y^b$. Note that these relationships are: (i) symmetric because when a cloud service can interoparate with another cloud service the latter can interoperate with the first one otherwise none of them could interoperate; (ii) not transitive, as there may exist a cloud service x^a that is interoperable with another service y^b using a protocol P while y^b interoperates with z^c using protocol P' but there may not exist a protocol such that allows x^a to interoperate with z^c.

Using these notations we will define two sets that are used in the definitions thereafter: *interoperable service set* and *interoperable types set*.

1. *Interoperable service set:* For a given cloud service $s^a \in CS_a$, this set is defined as the set of services in CS, that are interoperable with s^a:

$$S(s^a) = \{x^b | s^a \leftrightarrows x^b\} \tag{1}$$

   For example, when $S(s^a) = \{y^c, z^d\}$, it means that s^a is only interoperable with cloud services y^c and z^d.

2. *Interoperable types set:* For a given service type $t \in T$, this set is defined as the set of types of services in T that can interoperate with at least one cloud service of type t:

$$T_t = \{a | a \in T \wedge \exists x, y : x^t \leftrightarrows y^a\} \tag{2}$$

   Consequently, $T_a = \{b, c\}$ means that services of type a are only interoperable with services of types b and c.

In the following subsections, we define different metrics to score the interoperability of a cloud service s^a to provide an estimate of its current ability to function with other services, but also its capacity to extend or replace its functionality with other services in CS. Additionally, a balance among the types of functionalities that can extend the service is desired, as well as, high quality services to interoperate with. Considering these dimensions, an ideal set of metrics to measure the interoperability of a cloud service can be classified in two domains: quantitative metrics and qualitative metrics.

2.2.1   Quantitative Metrics

These metrics measure the capacity of a cloud service to interoperate with other services in terms of cardinality of the interoperable service set and their types. More precisely, these metrics measure the number of services and types compatible with s^a out of all the services and types under consideration. Such metrics reflect two particular aspects of a cloud service in multi-vendor environment: first, they measure the ability of a cloud service to operate with all available services and types in CS; second, they help to evaluate the complexity of replacing the services that are currently extending it.

Interoperable cloud service coverage: It measures the ability of the cloud service to interoperate with the highest number of services under consideration. In addition, the larger the number of interoperable cloud services, the larger the number of services that can be used in combination with it and, therefore, the larger the number of options available to extend its functionality. More formally, the interoperable cloud service coverage of a cloud service s^a is defined as the ratio of the number of services that are interoperable with s^a and the number of services belonging to types that are interoperable with services of type a.

$$\text{service}_{\text{cover}}(s^a) = \frac{|S(s^a)|}{|\bigcup_{t \in T_a} CS_t|}, \tag{3}$$

For example, let us suppose that s^a is an instance of a DBaaS type. We suppose that the only desirable interoperable service types with DBaaSs are monitoring tools and backup systems. In this case, the interoperable cloud service coverage of s^a should be proportional to the number of services belonging to the compatible types. The interoperable cloud service coverage of the cloud database s^a is the ratio of the number of services that are interoperable with it and the number of cloud services that are either monitoring tools or backup systems, because these are the only desirable interoperable types with DBaaSs. Note that: $\text{service}_{\text{cover}}(s^a) \in [0, 1]$.

Interoperable cloud service type coverage: It measures the ability of the cloud service to interoperate with the highest number of types from the services under consideration. More formally, cloud service type coverage is expressed as the ratio of the number of types of cloud services that are interoperable with the selected cloud service and the total number of types of cloud services that are interoperable with the type of the selected service:

$$\text{type}_{\text{cover}}(s^a) = \frac{|\{b|x^b \in S(s^a)\}|}{|T_a|} \tag{4}$$

For example, attaching a cloud monitoring tool and a backup tool to a compute instance adds two different functionalities to the latter: the possibility of being monitored and the possibility of recovering the state from a previous version. Note that: $\text{type}_{\text{cover}}(s^a) \in [0, 1]$.

Additionally, the relevance of being able to attach a specific type of service to a cloud service instance strongly depends on the requirements of the application that is going to use that service and it is also linked to the user's preferences. For instance, there may be a cloud storage system used to store very important information. The user knows that the load will be small but that data integrity is critical. Consequently, the possibility of being able to connect a backup system to the storage system becomes much more important than being able to monitor its load. We denote the preference of the user to extend the functionality of the selected cloud service type with an interoperable cloud service of type t as $w_t \in [0, 1]$. The user preferences are defined in such a way that for an specific type, all the user preferences of this type satisfy that: $\sum_{i=0}^{|T|} w_i = 1$. Using these user preferences, we define the weighted interoperable cloud service type coverage score that takes into account the relevance of each type:

$$wtype_{\text{cover}}(s^a) = \frac{\sum_{\{b | x^b \in S(s^a)\}} w_b}{\sum_{c \in T_a} w_c}, |T_a| > 0 \tag{5}$$

(5) computes the ratio of the sum of weights of the types that are interoperable with s^a and the sum of the weights of the types that are compatible with type a. When $\sum_{c \in T_a} w_c = 0$ then $wtype_{\text{cover}}(s^a) = 0$. Note that: $wtype_{\text{cover}}(s^a) \in [0, 1]$.

2.2.2 Qualitative Metrics

These metrics measure the capacity of cloud service to interoperate with other services taking into account the quality of the interoperable services and balance in their types. Such metrics reflect two abilities of a cloud service in multi-vendor environment: first, they help to estimate the quality of the enhancements that will be obtained by extending the functionalities of the selected cloud services. Second, they help to understand the distribution of cloud service types that can be attached to a cloud service and hence understand the variety of functionalities that can be added.

Interoperable cloud service quality: This metric measures the functional and/or non-functional quality of the cloud services that are interoperable with a given cloud service. There are many paradigms to assess the quality, q_x, of a cloud service x. These can be classified in three main categories based on the nature of the information used to generate the quality score: (i) based on user satisfaction [8]; (ii) based on the performance and quality of service (QoS) of the cloud services [3,4]; (iii) based on combined approaches that include user satisfaction and performance information into a single value [6]. Overall, q_x provides an estimate of the individual quality of the corresponding cloud service.

The interoperable cloud service quality is measured as the aggregation of the quality score of the services interoperable with the selected cloud services.

This aggregation can be done in different ways based on the desired evaluation of a cloud service. We define two such aggregation methods: average and weighted average.

Average Interoperable Cloud service quality: For a given service s^a, this metric is defined as the average of the quality scores of its interoperable services:

$$\text{quality}_{\text{avg}}(s^a) = \sum_{x \in S(s^a)} \frac{q_x}{|S(s^a)|} \tag{6}$$

A higher value of $quality_{\text{avg}}(s^a)$ indicates a higher average quality of services that are interoperable with s_a.

Weighted Average Interoperable Cloud service quality: For a given service s_a, it computes the average quality of the interoperable services with s_a weighed according to the importance of their types. It is computed in two steps: firstly, the average interoperability cloud service quality is computed for each type of the interoperable services with s_a:

$$\text{quality}^t_{\text{avg}}(s^a) = \sum_{x \in S(s^a) \cap \text{CS}_t} \frac{q_x}{|S(s^a) \cap \text{CS}_t|} \tag{7}$$

Secondly, the average value for each of the types are aggregated weighting them according to the user preferences w_i for selecting an specific type (defined in the previous subsection) as follows:

$$w\text{quality}_{\text{avg}}(s^a) = \sum_{t \in T_a} w_t \, \text{quality}^t_{\text{avg}}(s^a) \tag{8}$$

Note that $w\text{quality}_{\text{avg}}(s^a) \in [0, 1]$ because $w_t, q_x \in [0, 1]$ and $\sum_{i=0}^m w_i = 1$.

Other aggregation functions can be used apart from the *average* function. For example, the *maximum* value function would provide the score of the best service from each type. This function can be useful in situations in which one prefers to select the cloud service that is interoperable with the service that has the highest quality score. Similarly, *median* function can also be used instead of the *average* to provide a more robust value in the presence of outliers. The exploration of these alternative options in detail is beyond the scope of this paper.

Interoperable cloud service type entropy: This metric measures the variety among the types of the cloud services that a selected cloud service is interoperable with. The interoperable cloud service type entropy, H_t, helps to determine whether the services interoperable with s^a are skewed towards some cloud service types or not. This information can help to avoid situations in which the decision power is constrained by the number of interoperable services within each type. For example, let us assume that there are two cloud storage systems. One is interoperable with 9 backup tools and 1 log analyzer, the other is compatible with 5 back up tools and 5 log analyzers. Both cases would obtain the same scores when using (3), (4) or (5). However, having a high number of interoperable services in each type can be desirable to ensure sufficient flexibility to choose the services from, and reduce the lock-in risk. For example, let us assume one selects the former option and the single log analyzer becomes unusable for some reason. In this case, in order to use a log analyzer one should replace the selected cloud storage system by another one and go through a significant hassle and extra costs to carry out the replacement process.

H_t is computed in two steps. Firstly, we measure the conditional probability for a cloud service to be of type t given that it is connected to the selected cloud service s^a. This probability can be computed as follows:

$$p(s^a, t) = \frac{|S(s^a) \cap \mathrm{CS}_t|}{|S(s^a)|} \tag{9}$$

Secondly, the expression for information entropy is used to compute the skewness of the number of cloud services in each type using $p(s^a, t)$:

$$H_t(s^a) = - \sum_{t \in \{b | x^b \in S(s^a)\}} p(s^a, t) \log p(s^a, t) \tag{10}$$

The entropy metric helps to understand the qualitative distribution of the different types of functionalities that can extend the features of a particular cloud service. Evaluating the cloud service on the basis of its entropy can be useful to choose a service easy to migrate w.r.t. associated services in future. In the next section, we will evaluate these metrics to score different cloud services from the interoperability point of view.

3   Experiments

In this section, we evaluate the metrics described in Sect. 2 for a use case example and draw conclusions on the basis of extensive analysis. In this example, we assume that a company desires to deploy a Customer Relationship Management tool (CRM) in the cloud. This company has to choose, among others, a cloud database service to host the database of the system. In addition, it also requires to attach the database to a monitoring tool and to a log analyzer tool to keep track of the performance and reliability of the platform. In this way, the selection of a cloud database is conditioned not only by its interoperability with the CRM software but also by the tools that can be connected to it.

The experimental setup to test the proposed framework along with some numerical results is described in the following subsections.

3.1   Data Preparation

Data used in these experiments has been synthesized from publicly available data sources due to legal constraints. We explored different cloud providers' official websites to generate a dataset that mimics the actual interoperability of cloud services in the market. The steps followed to gather data are: (i) Collect the requirements of the CRM to be deployed to understand the requirements for the DB; (ii) Discover what cloud databases meet CRM's requirements; (iii) List relevant services that can be used along with the selected cloud services. In this case, we gather information about interoperable monitoring tools and log analyzers; (iv) Find performance metrics and quality scores for the individual cloud services providing monitoring tool and log analyzer functionalities.

The table in Fig. 1 lists a set of candidate cloud databases that are compatible with CRM, monitoring tools and log analyzers that are relevant, along with their quality score. The quality score used in this example is based on online user reviews. User reviews must be extracted from trusted datasources and sites that aggregate the opinion of many users. The interoperability among these cloud services is shown in Fig. 1, in which cloud services are nodes and edges connect pairs of services $\{a, b\}$ if and only if $a \leftrightarrows b$.

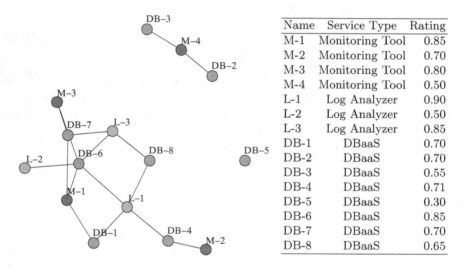

Name	Service Type	Rating
M-1	Monitoring Tool	0.85
M-2	Monitoring Tool	0.70
M-3	Monitoring Tool	0.80
M-4	Monitoring Tool	0.50
L-1	Log Analyzer	0.90
L-2	Log Analyzer	0.50
L-3	Log Analyzer	0.85
DB-1	DBaaS	0.70
DB-2	DBaaS	0.70
DB-3	DBaaS	0.55
DB-4	DBaaS	0.71
DB-5	DBaaS	0.30
DB-6	DBaaS	0.85
DB-7	DBaaS	0.70
DB-8	DBaaS	0.65

Fig. 1. The figure depicts the interoperability data of the cloud services used in the experiments and the table lists their types and scores.

3.2   Experimental Setup

The metrics proposed in Sect. 2 have been computed for each of the cloud databases included in this experiment with the following considerations:

- The set of cloud services listed in the 1st column of table included in Fig. 1 is considered CS in this experiment.
- The set of types T is composed by the types shown in the 2nd column of the table included in Fig. 1: cloud databases, monitoring tools and log analyzers.
- The values listed in the last column of table in Fig. 1 are used as the quality score q_x defined in Subsect. 2.2.2. Although inspired by real services, this values are synthetic to be compliant with terms of use and conditions of the official CSP websites.

Table 1. Values of the contextual interoperability metrics for each of the cloud databases included in our experiments.

	service$_{cover}$	type$_{cover}$	$w$type$_{cover}$	H_t	quality$_{avg}$	$w$quality$_{avg}$
DB-1	0.29	1.00*	1.00*	1.00*	0.88*	0.89*
DB-2	0.14	0.50	0.20	0.00	0.50	0.10
DB-3	0.14	0.50	0.20	0.00	0.50	0.10
DB-4	0.29	1.00*	1.00*	1.00*	0.80	0.86
DB-5	0.00	0.00	0.00	0.00	0.00	0.00
DB-6	0.71*	1.00*	1.00*	0.97	0.78	0.77
DB-7	0.43	1.00*	1.00*	0.92	0.83	0.85
DB-8	0.29	0.50	0.80	0.00	0.88*	0.70

- A pair of services is interoperable, i.e. $x^a \leftrightarrows y^b$, if and only if there is an edge in Fig. 1 connecting these two cloud services.
- The weights assigned to each service to extend the functionalities of the cloud DB are: Monitoring Tool preference 0.20 and Log Analyzer preference 0.80.

3.3   Numerical Results

The values of the metrics for each of the DBaaS in this scenario are presented in Table 1. From this table, we first highlight that DB-5 scores 0 in every metric used because it does not have any interoperable service among ones under consideration. The service$_{cover}$ value measures the percentage of all cloud services of a certain type that are interoperable to the corresponding cloud database. DB-6 obtains the highest value in this dimension because it is interoperable with maximum number of services as compared to other databases (see Fig. 1). According to type$_{cover}$ metric, it can be seen that there are 4 databases that can interoperate with at least one service of each of the two types (where this metric is 1). However, this estimate can provide only a small discriminative power in this scenario because there are only two different types of services. Though, $w$type$_{cover}$ provides higher scoring resolution because it is able to favor the cloud databases that are interoperable with the types of services that are more relevant to the user. Hence DB-8 scores better in this dimension and DB-2 and DB-3 score worse. Meanwhile, the interoperable cloud service type entropy, H_t, takes high values for the services that are connected with more than one type of service. That means that these services have a fairly balanced number of interoperable services of each type. Nonetheless, H_t provides valuable information such that the number of interoperable services per type is more balanced for DB-6 than for DB-7. Furthermore, the quality of the interoperable services has been evaluated using quality$_{avg}$ and $w$quality$_{avg}$. These metrics show interesting information. For instance, although DB-1 has a small service coverage, it is interoperable with the highest quality services. Analogously to the type diversity metrics analyzed above, $w$quality$_{avg}$

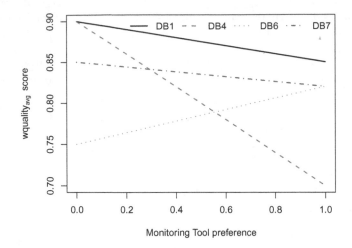

Fig. 2. Weighted average quality values as the user preference changes.

includes information about the user preferences on the types. Table 1 shows different cloud service rankings depending on the metric used.

Figure 2 shows the values for $wtype_{cover}$ when the user preference for the monitoring tools ranges from 0 to 1 (left to right) and the log analyzers preference from 1 to 0 (also from left to right). This figure includes the values of $wtype_{cover}$ for the top 4 cloud DBs from Table 1. It clearly shows the impact of the preference values over the final score of the cloud services. For example, DB-4 is more desirable than DB-6 when the preference for log analyzers is high and it is the way around when the preferred cloud service type are monitoring tools.

To sum up, these metrics measure new aspects that allow more accurate scoring of cloud services in multi-provider environments. With these, new opportunities arise in the process of choosing cloud services that fit better user requirements. Besides, the proposed metrics provide a systematic method to measure and compare the capacity of a cloud service to interoperate with other services in the same digital ecosystem.

4   Conclusion and Future Work

The evolving cloud service marketplace is witness of the lack of metrics and methods to compare and assess cloud services and to assist complex decision making processes for the actors involved in developing and deploying software for complex e-business scenarios. In this paper, we have made the first attempt to address the problem of scoring the interoperablity among cloud services in a multi-provider environment. For this, we provide a set of metrics that are not computed based on the particular characteristics of a cloud service, but on its relationship with the community of cloud services in its digital ecosystem. With this work, we aim at establishing a new basis for dealing with cloud service interoperability information.

The metrics presented in this paper measure the relationship of one cloud service with its digital community based on its compatibility. This is a first necessary step to examine how these metrics can be incorporated to contribute to the analysis of cloud service interoperability and to avoid vendor lock-in in multi-cloud environments.

Acknowledgment. This work is partially supported by Secretaria de Universitats i Recerca of Generalitat de Catalunya (2014DI031) and conducted as a part of the MUSA project (Grant Agreement 644429) funded by the European Commission within call H2020-ICT-2014-1. Josep L. Larriba-Pey also thanks the Ministry of Economy and Competitivity of Spain and Generalitat de Catalunya, for grant numbers TIN2013-47008-R and SGR2014-890 respectively.

References

1. Topology and orchestration specification for cloud applications version 1.0. Technical report, OASIS Standard, Nov 2013
2. Gupta, S., Muntes-Mulero, V., Matthews, P., Dominiak, J., Omerovic, A., Aranda, J., Seycek, S.: Risk-driven framework for decision support in cloud service selection. In: 2015 15th IEEE/ACM International Symposium on Cluster, Cloud and Grid Computing (CCGrid), pp. 545–554, May 2015
3. Han, S., Hassan, M., Yoon, C., Huh, E.: Efficient service recommendation system for cloud computing market. In: Proceedings of the 2nd International Conference on Interaction Sciences: Information Technology, Culture and Human, ICIS 2009, pp. 839–845. ACM, New York (2009)
4. Liu, F., Tong, J., Mao, J., Bohn, R., Messina, J., Badger, L., Leaf, D.: Nist cloud computing reference architecture. NIST Spec. Publ. **500**, 292 (2011)
5. Muntes-Mulero, V., Matthews, P., Omerovic, A., Gunka, A.: Eliciting risk, quality and cost aspects in multi-cloud environments. In: The Fourth International Conference on Cloud Computing, GRIDs, and Virtualization, CLOUD COMPUTING (2013)
6. Lie, Q., Yan, W., Orgun, M.A.: Cloud service selection based on the aggregation of user feedback and quantitative performance assessment. In: 2013 IEEE International Conference on Services Computing (SCC), pp. 152–159, June 2013
7. Sheng, Q.Z., Qiao, X., Vasilakos, A., Szabo, C., Bourne, S., Xu, X.: Web services composition: A decadés overview. Inf. Sci. **280**, 218–238 (2014)
8. ur Rehman, Z., Hussain, O.K., Hussain, F.K.: Iaas cloud selection using MCDM methods. In: IEEE Ninth International Conference on e-Business Engineering (ICEBE), pp. 246–251, Sept 2012
9. ur Rehman, Z., Hussain, O.K., Chang, E., Dillon, T.: Decision-making framework for user-based inter-cloud service migration. Electronic Commerce Research and Applications (2015)
10. Uriarte, R.B., Tsaftaris, S., Tiezzi, F.: Service clustering for autonomic clouds using random forest. In: 2015 15th IEEE/ACM International Symposium on Cluster, Cloud and Grid Computing (CCGrid), pp. 515–524, May 2015
11. Viveca, R., Meulen, W.: Gartner says worldwide public cloud services market is forecast to reach $204 billion in 2016. Technical report, Gartner Press Release (2016)

Handling Branched Web Service Composition with a QoS-Aware Graph-Based Method

Alexandre Sawczuk da Silva[1(✉)], Hui Ma[1],
Mengjie Zhang[1], and Sven Hartmann[2]

[1] School of Engineering and Computer Science, Victoria University of Wellington,
PO Box 600, Wellington 6140, New Zealand
{sawczualex,hui.ma,mengjie.zhang}@ecs.vuw.ac.nz
[2] Department of Informatics, Clausthal University of Technology,
Julius-Albert-Strasse 4, 38678 Clausthal-Zellerfeld, Germany
sven.hartmann@tu-clausthal.de

Abstract. The concept of Service-Oriented Architecture, where individual services can be combined to accomplish more complex tasks, provides a flexible and reusable approach to application development. Their composition can be performed manually, however doing so may prove to be challenging if many service alternatives with differing qualities are available. Evolutionary Computation (EC) techniques have been employed successfully to tackle this problem, especially Genetic Programming (GP), since it is capable of encoding conditional constraints on the composition's execution paths. While compositions can naturally be represented as Directed Acyclic Graphs (DAGs), GP needs to encode candidates as trees, which may pose conversion difficulties. To address that, this work proposes an extension to an existing EC approach that represents solutions directly as DAGs. The tree-based and extended graph-based composition approaches are compared, showing significant gains in execution time when using graphs, sometimes up to two orders of magnitude. The quality levels of the solutions produced, however, are somewhat higher for the tree-based approach. This, in addition to a convergence test, shows that the genetic operators employed by the graph-based approach can be potentially improved. Nevertheless, the extended graph-based approach is shown to be capable of handling compositions with multiple conditional constraints, which is not possible when using the tree-based approach.

Keywords: Web service composition · QoS optimisation · Conditional branching · Evolutionary computing · Graph representation

1 Introduction

In recent years, the concept of Service-Oriented Architecture (SOA) has become increasingly popular [10]. According to SOA, software systems should be organised into independent functionality modules known as *Web services* [2], which

© Springer International Publishing AG 2017
D. Bridge and H. Stuckenschmidt (Eds.): EC-Web 2016, LNBIP 278, pp. 154–169, 2017.
DOI: 10.1007/978-3-319-53676-7_12

are accessible over a network and employed in conjunction to fulfil the overall system's objectives. In other words, SOA encourages the idea of *Web service composition*, where existing services are reused to create new applications instead of re-implementing functionality that is already available. Such compositions can be performed manually, however this process may become time-consuming as the number of relevant candidate services increases. Thus, the ability to perform Web service compositions in a fully automated manner has become an important area of research in the field of service computing [8].

The process of Web service composition is challenging for three reasons: firstly, the *functional correctness* of service composition solutions must be ensured, i.e. the outputs and inputs of atomic services in the solution must be connected in a way that is actually executable; secondly, the creation of compositions with multiple execution *branches* must be supported, so that applications incorporating conditional constraints can be produced to adapt to dynamic changes of the environment [16]; thirdly, the *quality* (e.g. reliability, availability) of the candidate Web services to be included in composite services must be examined, so that the composite services can achieve the best possible qualities. Several automated approaches for Web service composition exist, but the great majority of them fails to address these three dimensions simultaneously. For instance, AI planning approaches focus on ensuring interaction correctness and branch support [14], while Evolutionary Computation (EC) techniques focus on interaction correctness and Quality of Service (QoS) measures [15].

Recently, a genetic programming (GP) approach was proposed to address these three composition dimensions simultaneously [13], but the tree representation used in this work may lead to situations in which the same service appears multiple times throughout a candidate tree, making it difficult to ensure the functional correctness of the composition solution (as further explained in Sect. 2). If solutions are represented as Directed Acyclic Graphs (DAGs), on the other hand, it becomes much easier to verify the correctness of the connections between component services in the composition. A DAG representation for evolutionary Web service composition was proposed in [12], however that work does not allow for the creation of branches. Thus, the objective of this work is to propose an extension to that graph-based Evolutionary Computation Web service composition approach so that it is capable of simultaneously addressing the three dimensions discussed above. The remainder of this paper is organised as follows. Section 2 describes the composition problem and existing works. Section 3 introduces the proposed graph-based composition approach. Section 4 outlines the experiments conducted to compare the performance of graph-based and tree-based approaches. Section 5 concludes the paper.

2 Background

2.1 Existing Works

From the existing EC techniques, GP is perhaps the most flexible within the domain of Web service composition, since it allows a variable amount of services to be encoded into a solution in multiple configurations that are dictated by the chosen composition constructs. GP represents each solution as a tree, using one of two possible representations: the *input-as-root* representation [11, 13], where composition constructs (e.g. parallel, sequence) are encoded as non-terminal nodes of the tree and atomic Web services as terminal nodes, and the *output-as-root* representation [5], where the root node of the tree represents the composition output, the other non-terminal nodes are atomic Web services, and all of the terminal nodes represent the composition input.

The advantage of these two representations is that their crossover and mutation operations are relatively simple to implement, since modifications are restricted to certain subtrees. However, both of these approaches also present some disadvantages. The input-as-root representation using strongly-typed GP has the disadvantage of duplicating Web services throughout the tree, which means that modifying a part of the tree that contains a duplicated service may remove some fundamental connections to this service, thus breaking the correctness of the solution. The output-as-root representation, on the other hand, does not have this type of modification problem; however, it does not support conditional branching. That is because the set of composition outputs is represented as the root of the tree, meaning that there is only one set of outputs can be represented at a time. Another disadvantage of these techniques is that a graph representation must be generated first, and subsequently translated into a tree representation to ensure the functional correctness of trees.

Mabu, Hirasawa and Hu [6] introduce a technique for the evolution of graph-based candidates, called Genetic Network Programming (GNP). GNP has a fixed number of nodes within its structure, categorised either as processing nodes (responsible for processing data) or as judgment nodes (perform conditional branching decisions), and works by evolving the connections between these fixed nodes. While this approach has the advantage of using simple genetic operations, the number of nodes and outgoing edges per node must be fixed throughout the evolutionary process. Other graph-based evolutionary techniques [1, 9] do allow for flexible topologies, but they do not cater for the structural constraints that must be observed in a service composition.

To address these limitations, Silva, Ma and Zhang [12] introduce GraphEvol, an EC technique where Web service composition solutions are represented as Directed Acyclic Graphs (DAGs). However, GraphEvol neither supports QoS-aware composition nor allows conditional constraints to be used. Thus, in this paper we will employ a DAG representation to perform QoS-aware service composition with conditional constraints. While the work of Wang et al. [16] does take conditional constraints into account, it does not allow for multiple possible outputs (i.e. separate branches that are not merged) and it is not QoS-aware.

2.2   A Motivating Example

A typical Web service composition scenario is that of an online book shopping system, adapted from [13] and shown in Fig. 1. The objective of this system, constructed using existing Web services, is to allow a customer to purchase a

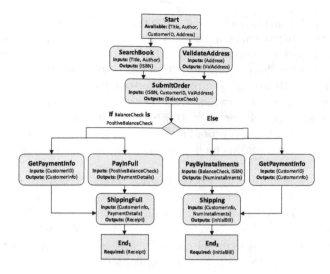

Fig. 1. A Web service composition for an online book shopping system [13].

book using different methods of payment according to their account balance. Namely, if the customer has enough money to pay for the selected book, then they would like to pay for it in full; otherwise, they would like to pay for it in instalments. The idea is to construct this system in a fully automated manner, meaning that the services to be used, the connections between these services, and the overall flow of the system are determined automatically. When requesting the creation of a system with the desired functionality, the composition request provided consists of the overall inputs required by the system (e.g. book title, author, customer ID, address), conditional constraints (e.g. the customer's account balance is less than the price of the chosen book), and the potential outputs produced by the system (e.g. a receipt, if the customer paid in full, or an initial bill, if the customer is paying in instalments).

2.3   Problem Formalization

A *Web service* S is represented using a functional description that includes an input concept I and an output concept O specifying what operation the service will perform, a categorical description of inter-related concepts specifying the service operation according to a common terminology $\mathcal{T}$ in the application area, and a quality of service (QoS) description of non-functional properties such as response time and cost. For the categorical description, an ontology with definitions of "concepts" and the relationships between them must be specified. In previous works [4] a terminology for service ontology using description logic has been defined. A *terminology* is a finite set $\mathcal{T}$ of assertions of the form $C_1 \sqsubseteq C_2$ with concepts C_1 and C_2 as defined in [4].

A *service repository* $\mathcal{D}$ consists of a finite collection of atomic services $s_i, i \in \{1, n\}$ together with a service terminology $\mathcal{T}$. A *service request* R is normally defined with a pair (I_R, O_R), where I_R is the input concept that users provide and O_R specifies the output concept that users require. In this paper, we consider that users may prefer different outputs depending on some condition [16]. In this case, we specify the output as $(c?O_R, O_{R'})$, meaning that if the value of the condition c is TRUE, then it produces output O_R, otherwise $O_{R'}$. A service request can be represented as two special services, a start service $s_0 = (\emptyset, I_R)$ and an end service $s_e = \{O_R, \emptyset\}$ or $s_{e'} = \{O_{R'}, \emptyset\}$, when conditional constraints are used. Each output has a given probability of being produced, which is calculated through statistical analysis on the behaviour of the service(s) used to satisfy the request.

Services can be composed by process expressions [4]. The set of *process expressions* over a repository $\mathcal{D}$ is the smallest set $\mathcal{P}$ containing all *service constructs* that is closed under sequential composition construct $\cdot$, parallel construct $\|$, choice $+$ and iteration $*$. That is, whenever $s_i, s_j \in \mathcal{P}$ hold, then all $s_i \cdot s_j$, $s_i \| s_j$, $s_i + s_j$ and s_i* are process expressions in $\mathcal{P}$, with s_i and s_j are component services. A *service composition* is a process expression with component services s_i, each of which associating with an input and output type I_i and O_i. For example $s_1 \cdot (s_2 \| s_3)$ is a service composition, meaning that s_1 processed first followed with s_2, s_3 processed in parallel.

We need to ensure that composite services are *feasible*, i.e. *functionally correct*. In particular, when two services are composed by a sequential construct, i.e., $s_i \cdot s_j$, we need to ensure that service s_j matches s_i. A service s_j *fully matches* another service s_i if and only if O_i subsumes concept I_j, i.e., $O_i \sqsubseteq I_j$. A service s_j *partially matches* another service s_i if $O_i \sqcap I_j \sqsubseteq \bot$. A service composition $\mathcal{S}$ is a *feasible* solution for the service request R if the following conditions are satisfied:

- All the inputs needed by the composition can be provided by the service request, i.e. $I_R \sqsubseteq I_\mathcal{S}$;
- All the required outputs can be provided by the composition, i.e. $O_\mathcal{S} \sqsubseteq O_R$ if c is TRUE and $O_\mathcal{S} \sqsubseteq O_{R'}$, otherwise;
- For each component service s_j its input I_j should be provided by services s_i that were executed before it, i.e. the union of the output of all the services s_i is a sub-concept of I_j.

Service compositions can be naturally represented as Directed Acyclic Graphs (DAGs) [12]. Given a service repository $\mathcal{D}$ and a service request R, the service composition problem is to find a directed acyclic graph $G = \{V, E\}$, where V is the set of vertices and E is the set of edges of the graph, with one starting service s_0 and end service s_e and a set of intermediate vertices $\{V_1, \ldots, V_m\}$, which represent atomic services selected from the service repository.

Service compositions can be also represented as trees, with intermediate nodes representing composition constructs and terminal nodes representing atomic services [13]. Each intermediate node is actually a composite service itself, with child nodes arranged according to the composition construct it represents. Therefore, to check the functional correctness of service composition represented as a tree we just need to check that each of the component services, including atomic services, satisfies the following: its input concept must be subsumed by the union of the properties of the input from the service requests and the properties produced by its preceding nodes.

2.4   QoS Model and Composition Constructs

In addition to the functional aspects of Web service composition, the goodness of the services included in a solution also plays a part in the creation of a composite system. This non-functional set of attributes is known as Quality of Service (QoS) [7], and it measures characteristics that are desirable in a service from a customer's point of view. In this work, four QoS attributes are considered: Time (T), which measures the response time of a service once it has been invoked, Cost (C), which specifies the financial cost of using a given service, Availability (A), which measures the likelihood of a service being available at invocation time, and Reliability (R), which is the likelihood of a service responding appropriately when invoked. The existing languages for Web service composition (e.g. BPEL4WS [17]) use certain constructs to control the flow of the resulting systems with regards to input satisfaction.

2.4.1   Sequence Construct

For a composite service $S = s_1 \cdot ...s_n \cdot ...s_m$ services are chained sequentially, so that the outputs of a preceding service are used to satisfy the inputs of a subsequent service. The total cost C and time T of the services are calculated by adding the quality values of its individual services, and the total availability and reliability by multiplying them ($T = \sum_{n=1}^{m} t_n, C = \sum_{n=1}^{m} c_n, A = \prod_{n=1}^{m} a_n, R = \prod_{n=1}^{m} r_n$).

2.4.2   Parallel Construct

For a composite service $S = s_1||...s_n||...s_m$ services are executed in parallel, so their inputs are independently matched and their outputs are independently produced. The availability, reliability and cost are calculated the same way as they are in the sequence construct, and the total time is determined by identifying the service with the longest execution time ($T = MAX\{t_n|n \in \{1,...,m\}\}$).

2.4.3   Choice Construct

For a composite service $S = s_1 + ...s_n + ...s_m$ only one service path is executed, depending on whether the value of its associated conditional constraint is met at runtime. For S, all overall QoS attributes are calculated as a weighted sum of the services from each individual path, where each weight corresponds to the probability of that path being chosen during runtime ($T = \sum_{n=1}^{m} p_n t_n$, $C = \sum_{n=1}^{m} p_n c_n$, $A = \sum_{n=1}^{m} p_n a_n$, $R = \sum_{n=1}^{m} p_n r_n$). These weights add up to 1.

3   Proposed Approach

The approach proposed in this work allows for the representation of composite services with multiple execution branches, depending on conditional constraints specified in the composition request. However, nodes with multiple children and/or multiple parents can also be included when necessary, meaning that Directed Acyclic Graphs (DAGs) are the basis of this representation. Algorithm 1 describes the general procedure followed by GraphEvol, and the following subsections explain the ways in which this basic technique was extended to allow for conditional branching.

Algorithm 1. Steps of the GraphEvol technique [12].

---

1   Initialise the population using the graph building Algorithm 2.
2   Evaluate the fitness of the initialised population.
    while *max. generations not met* **do**
3 | Select the fittest graph candidates for reproduction.
4 | Perform mutation and crossover on the selected candidates, generating offspring.
5 | Evaluate the fitness of the new graph individuals.
6 | Replace the lowest-fitness individuals in the population with the new graph individuals.

3.1   Graph Building Algorithm

Candidate compositions S are created using the *buildGraph* procedure shown in Algorithm 2. This procedure requires *InputNode* I_r, which is the root of the composition's request tree, a list of *relevant* candidate services from the service repository D, and optionally a *candMap*. The *candMap* is used for the crossover procedure only, so it will be discussed later. Given these inputs, the algorithm proceeds to connect nodes to the graph, one at a time, until a complete solution S is found. As explained earlier, the resulting composition will have several independent branches, thus the recursive procedure *buildBranch* has been created to handle each part of the composition. After connecting the *start* service s_0 to the graph, we execute *buildBranch* providing the first task it should achieve (i.e. *TaskNode*, which initially will be a conditional branching node – i.e. a c node), a list of candidates *candList* that contains services that are executable/reachable using the *start* node outputs, the partially built graph G, and other relevant data structures. Once the *buildBranch* procedure has finished executing, the graph G representing the composition S will be completed. The algorithm used for construction creates graphs from the *start* node to the *end* nodes s_e and $s_{e'}$ in order to prevent cycles from forming, but this may lead to *dangling* nodes, which are nodes that do not have any outgoing edges despite not being *end* nodes. These are redundant parts of the solution, and thus they must be removed once G is built. Finally, the creation of the new candidate graph is finished.

Algorithm 2 also describes the *connectNode* function, used for adding a node to an existing graph. In addition to adding the given node n to G, and connecting it using the edges provided in the *connections* list, this function also checks if the current *TaskNode* objective has been reached. If the *TaskNode* represents a conditional node c, we check that we are now capable of producing both the values required by the *if* case (O_r) and by the *else* case $(O_{r'})$ when using the service we have just connected. On the other hand, if the *TaskNode* represents the *end* of a branch, we check that the list *allInputs* of potential inputs contain all values necessary to satisfy the inputs for the *end* node – either s_e or $s_{e'}$.

Algorithm 2. Procedures for building a new candidate graph and for connecting a particular node to the graph [12].

```
 1: Procedure buildGraph(InputNode, relevant, candMap)
 2:     start.outputs ← {InputNode.values};
 3:     TaskNode ← InputNode.child;
 4:     G.edges ← {};
 5:     G.nodes ← {};
 6:     allInputs ← {};
 7:     connections ← {};
 8:     connectNode(start, connections, G, allInputs, TaskNode);
 9:     allowedAncestors ← {start};
10:     if candMap is null then
11:         candList ← findCands(start, allowedAncestors, relevant);
12:     else
13:         candList ← {node|(start, node) ∈ candMap};
14:     buildBranch(TaskNode, candList, allInputs, G, relevant, allowedAncestors,
             candMap);
15:     removeDangling(G);
16:     return G;

17: Function connectNode(n, connections, G, allInputs, TaskNode)
18:     n.objective ← TaskNode;
19:     G.nodes ← G.nodes ∪ {n};
20:     G.edges ← G.edges ∪ connections;
21:     if TaskNode is ConditionNode then
22:         if |n.outputs| > 1 then
23:             return ((TaskNode.general ⊑ n.outputs.general ∧ TaskNode.specific ⊑
                     n.outputs.specific), n);
24:         else
25:             return (false , n);
26:     else
27:         return (TaskNode.outputs ⊑ allInputs, n);
```

Algorithm 3 shows the *buildBranch* procedure, which recursively creates the branched Web service composition. Given a *TaskNode*, this procedure repeatedly adds services to the graph G, until the *TaskNode* goal has been reached. More specifically, nodes from the *candList* are considered for addition. A candidate *cand* is randomly chosen from *candList*, and it is connected to the graph (using the *connectNode*) procedure if all of its inputs can be matched by the *ancestor* outputs (i.e. the outputs of nodes already present in that particular execution branch). The set of services in *connections*, which are used to connect *cand* to G, is a minimal set, meaning that the output of these services matches all the inputs of *cand*, but if any connection is removed from the set that is no longer the case. After each *cand* service is connected to G, the *candList* is updated to contain any services that have now become executable due to the outputs of *cand*, and to exclude *cand*. Once the *TaskNode* goal has been reached, the *connectTaskNode* procedure is called to finish the construction of that branch, either by connecting an *end* node to it (s_e or $s_{e'}$) or by further splitting the branch according to a new *TaskNode* condition c. In case of the latter, *connectTaskNode* will invoke the *buildBranch* procedure again.

Algorithm 3. Indirectly recursive procedure for building one of the branches of the new candidate graph.

```
 1: Procedure
      buildBranch(TaskNode, candList, allInputs, G, relevant, allowedAncestors, candMap)
 2:     goalReached ← false ;
 3:     connResult;
 4:     while ¬goalReached do
 5:        found ← false ;
 6:        foreach cand ∈ candList do
 7:           connections ← {};
 8:           ancestors ← {x.outputs|x ∈ G ∧ x ∈ allowedAncestors};
 9:           if cand.inputs ⊑ ancestors then
10:              connections ← connections ∪ {x ← minimal(ancestors)};
11:              found ← true ;
12:           if found then
13:              connResult ← connectNode(cand, connections, G,
                    allInputs, TaskNode);
14:              goalReached ← connResult[0];
15:              allowedAncestors ← allowedAncestors ∪ {cand};
16:              if candMap is null then
17:                 candList ← candList∪ findCands(cand, allowedAncestors,
                       relevant);
18:              else
19:                 candList ← candList ∪ {node|(cand, node) ∈ candMap};
20:              break;
21:           candList ← candList − {cand}
22:     connectTaskNode(TaskNode, connResult, G,
           allowedAncestors, candList, candMap);
```

As previously explained, Algorithm 4 is responsible for finishing the construction of a given execution branch, according to one of two scenarios. In the first scenario, the *TaskNode* reached is a conditional node c, meaning that the branch will be further split into an *if-and-else* structure. In this case, the *TaskNode* is added to G, connected through the previously added service in *connResult*[1] (i.e. the service that matched the outputs required for the condition to be checked). Since the branching occurs based on the values produced by $connResult[1]$, the probabilities of producing these different output possibilities are copied from this service. Then, the $buildBranch$ procedure is invoked twice more, once for the if branch and once for the $else$ branch, providing the appropriate children of $TaskNode$ to the next construction stages. In the second scenario, the $TaskNode$ reached is an output node, meaning that the branch leads to an end node without any further splitting (s_e or s'_e). In this case, the $TaskNode$ is simply connected to G, using a minimal set of services already in the graph which produce all the outputs required by this end node.

Algorithm 4. Procedure for finishing construction of a branch, splitting it further in case another condition exists.

```
1: Procedure
   connectTaskNode(TaskNode, connResult, G, allowedAncestors, candList, candMap)
2:     if TaskNode is ConditionalNode then
3:         TaskGoal.probs ← connResult[1].probs;
4:         G.nodes ← G.nodes ∪ {TaskNode};
5:         G.edges ← G.edges ∪ {connResult[1] → TaskNode};
6:         allowedAncestors ← allowedAncestors ∪ {TaskNode};
7:         connections ← {};
8:         if candMap is null then
9:             candList ← candList∪ findCands(TaskNode, allowedAncestors,
                   relevant);
10:        else
11:            candList ← candList ∪ {node|(TaskNode, node) ∈ candMap};
12:        allInputs ← {};
13:        ifChild ← TaskNode.ifChild;
14:        if ifChild is OutputNode then
15:            allInputs ← {x.outputs|x ∈ allowedAncestors};
16:        buildBranch(ifChild, candList, allInputs,
               G, relevant, allowedAncestors, candMap);
17:        elseChild ← TaskNode.elseChild;
18:        if elseChild is OutputNode then
19:            allInputs ← {x.outputs|x ∈ allowedAncestors};
20:        buildBranch(elseChild, candList, allInputs,
               G, relevant, allowedAncestors, candMap);
21:    else
22:        ancestors ← {x.outputs|x ∈ G ∧ x ∈ allowedAncestors};
23:        connections ← {x → TaskNode |x ∈ minimal(ancestors)};
24:        G.nodes ← G.nodes ∪ {TaskNode};
25:        G.edges ← G.edges ∪ connections;
```

3.2 Mutation and Crossover

The procedures for performing the mutation and crossover operations are shown in Algorithm 5. The general idea behind the *mutation* procedure is to modify a part of the original graph G, but maintain the rest of the graph unchanged. In order to do so, a node n is initially selected as the mutation point, provided that it is not an end node (s_e or $s_{e'}$) or a *condition* node c. If this node is the *start* node (s_0), an entirely new candidate graph is constructed; otherwise, all nodes whose input satisfaction depends upon node n are removed from G, and so are any subsequent splits of that branch. The construction of this partially-built graph is then finished by invoking the *buildBranch* procedure and providing the original *TaskNode* (*n*'s objective) and appropriate data structures to it. The *allowedAncestors*, the *candList*, and *allInputs* are calculated based on the remaining nodes of *G*. The mutation operator was designed in this way so that it allows for variations of the original candidate, at the same time maintaining the correctness of the connections between services.

In the case of *crossover*, the general idea is to reuse connection patterns from two existing candidates G_1 and G_2 in order to create a new child candidate that combines elements from these two parents. In order to do so, the original connections of G_1 and G_2 are abstracted into a map called *candMap*. This map can be queried to determine all the connections (from both parents) that can be made starting from a given node *x*. After having assembled this map, the *buildGraph* procedure is invoked to create a child candidate. The difference is that the addition of candidates to the *candList* is done by querying the *candMap* to determine which services could be reached from the current node according to the connection patterns in the original parents. One of the advantages of this crossover implementation is that it allows for an operation that reuses connection information from both parents. Additionally, this operation can be executed using the existing graph-building Algorithm 2 with minimal changes.

Algorithm 5. Procedures for performing mutation and crossover on graph candidates [12].

```
 1:  Procedure mutation(G, InputNode, relevant)
 2:      n ← selectNode(G);
 3:      if n is start then
 4:          return buildGraph(InputNode, relevant, null);
 5:      else
 6:          TaskNode ← n.objective;
 7:          removeNodes(n);
 8:          allInputs ← {};
 9:          candList ← {};
10:          allowedAncestors ← {};
11:          foreach node ∈ G.nodes do
12:              allowedAncestors ← allowedAncestors ∪ {node};
13:          foreach node ∈ G.nodes do
14:              candList ← candList∪ findCands(node, allowedAncestors, relevant);
15:          if TaskNode is OutputNode then
16:              allInputs ← {x.outputs|x ∈ allowedAncestors};
17:          return buildBranch(TaskNode, candList, allInputs,
                 G, relevant, allowedAncestors, null);

18:  Procedure crossover(G₁, G₂, InputNode, relevant)
19:      candMap ← {(x, y)|x → y ∈ G₁.edges};
20:      candMap ← candMap ∪ {(x, y)|x → y ∈ G₂.edges};
21:      return buildGraph(InputNode, relevant, candMap);
```

3.3 Fitness Function

The fitness function is used for optimising solutions according to their overall QoS, and was based on the function shown in [13]. This function measures the overall quality of a composition candidate by performing a weighted sum of the overall QoS attributes of a given candidate:

$$fitness_i = w_1A_i + w_2R_i + w_3(1 - T_i) + w_4(1 - C_i) \tag{1}$$

where $\sum_{k=1}^{4} w_k = 1$.

This function produces values in the range [0,1], where a fitness of 1 means the best quality. Because this is a maximising function, the Time T_i and cost C_i are offset by 1 in the formula, so that higher scores correspond to better qualities for these attributes as well. The overall quality attributes A_i, R_i, T_i, and C_i of a composition $\mathcal{S}_i$ are normalised, with the upper bound of T_i and C_i multiplied by the total number of services in the repository [5].

4   Evaluation

Experiments were conducted to compare the performance of our approach to that of another composition technique that uses a GP tree representation to encode branching constraints [13], as no other existing evolutionary composition approaches support the creation of compositions with branches and multiple sets of outputs. This tree-based approach follows the input-as-root representation discussed in Sect. 2. The datasets employed in this comparison were based on those proposed in [13], since they contain composition requests that require one conditional constraint c. The initial sets were extended to make the problem more complex by replicating each original service in the repository ten times, to more thoroughly test the scalability of the new approach. The replicated services were then assigned randomly generated QoS values that were within the original ranges for each quality attribute. A new dataset was created to measure the performance of our technique when addressing more complex composition requests. More specifically, this dataset was based on task 1 of the dataset presented in [13], with two changes: firstly, manually-generated services whose composition results in a solution with fitness 1 (i.e. an artificial optimum) were added; secondly, the composition request was modified to require three conditional constraints instead of one, as before.

Tests were executed on a personal computer with an Intel Core i7-4770 CPU (3.4 GHz), and 8 GB RAM. In total, four different tests were executed, three of them being comparisons and the fourth being an experiment on the behaviour of our approach only. The parameters for all approaches were based on those proposed by Koza [3]. 30 independent runs were conducted for each approach, with a population size of 500 during 51 generations. The crossover probability was set to 0.8, and mutation and reproduction probabilities were set to 0.1 each. No elitism was used, and tournament selection with a tournament size of 2 was employed. Finally, all fitness function weights were set to 0.25, indicating that all quality attributes are considered equally important by the user requesting the composition. The results of the comparison between the two composition techniques using the datasets from [13] are shown in Table 1, where the first column lists the dataset used and its number of atomic services, the second column displays the mean time with standard deviation for executing our graph-based approach, the third column shows the mean fitness and standard deviation of the best solution found by our graph-based approach, and the fourth and fifth columns show the corresponding time and fitness values for the tree-based approach. Time values, including their standard deviations, are rounded to 1 decimal point of precision; fitness values and their standard deviations are rounded to 2 decimal points. Wilcoxon signed-ranked tests at 95% confidence level were run where possible to ascertain whether the differences in time and fitness values for a given dataset are significant, with $\downarrow$ denoting significantly lower values and $\uparrow$ denoting significantly higher values.

Table 1. Comparison results.

Set (size)	Graph-based		Tree-based	
	Avg. time (s)	Avg. fitness	Avg. time (s)	Avg. fitness
1(1738)	$11.2 \pm 1.5 \downarrow$	0.76 ± 0.02	235.2 ± 52.8	$0.85 \pm 0.01 \uparrow$
2(6138)	$35.0 \pm 3.3 \downarrow$	$0.67 \pm 0.01 \uparrow$	609.3 ± 112.7	0.65 ± 0.00
3(6644)	$19.0 \pm 1.0 \downarrow$	0.72 ± 0.01	2264.5 ± 296.2	$0.74 \pm 0.02 \uparrow$
4(11451)	$49.1 \pm 1.6 \downarrow$	0.56 ± 0.01	900.6 ± 138.2	$0.77 \pm 0.08 \uparrow$
5(11990)	$34.9 \pm 1.3 \downarrow$	0.81 ± 0.02	2680.7 ± 217.8	$0.83 \pm 0.01 \uparrow$
6(24178)	$140.7 \pm 21.8 \downarrow$	$0.77 \pm 0.02 \uparrow$	19772.2 ± 2142.7	0.76 ± 0.02
7(45243)	$345.4 \pm 55.5 \downarrow$	0.79 ± 0.02	24467.1 ± 5482.4	$0.90 \pm 0.03 \uparrow$
8(89309)	$522.1 \pm 94.5 \downarrow$	0.82 ± 0.00	51850.3 ± 5768.2	0.82 ± 0.04

As expected, the execution times of our graph-based approach are significantly lower than those of the tree-based approach for all datasets. Surprisingly, the performance gains of the graph-based method are more pronounced as the size of the dataset grows, culminating into a difference of two orders of magnitude for dataset 8. These results are extremely encouraging, because they demonstrate that representing solutions directly as a DAG facilitates the enforcement of correct output-input connections between services, which in turn translates to lower execution costs. With regards to the quality of solutions, results show that the fitness of the tree-based solutions is slightly higher for the majority of datasets, but occasionally the quality of the solutions produced using the graph-based approach is superior (datasets 2 and 6). These results indicate a trade-off between the techniques, depending on whether the objective is to produce solutions using a lower execution time or to focus on the quality of these solutions instead. However, the rate of improvement for these two aspects should also be taken into consideration when comparing techniques. Namely, while the tree-based approach may result in solutions with a quality gain of up to 20% (for dataset 4), the execution time required by the graph-based approach may be as little as 1% (for datasets 5 to 8) of that required by the tree-based approach.

Thus, our proposed graph-based approach is a viable Web service composition alternative.

A convergence test was also conducted using the variation of dataset 1 described earlier. Since the task associated with this new dataset requires the creation of a composition that takes multiple conditional constraints into account, it can only be tested against our graph-based approach (recall that the tree-based approach cannot handle more than one conditional constraint [13]). Therefore, the objective of this test is twofold: to demonstrate that the graph-based approach can indeed handle composition requests with multiple conditional constraints, and to examine the convergence of this approach. The results of this test are shown in Fig. 2, where the mean fitness values calculated over 30 independent runs have been plotted for each generation. The artificial optimum is known to correspond to the value 1 when evaluated using the fitness function, which means that solutions with a value close to 1 are likely to be close to this optimum. The plotted mean values clearly show that the most significant fitness improvements occur between generations 0 and 30, with the remaining generations performing smaller improvements that eventually lead to the convergence of the population. Interestingly, the fitness values between generations 40 and 50 remain constant at 0.88, without the slightest variation. This suggests that improvements to these operators could be performed to enhance the search strategies when using a DAG representation. Nevertheless, these results show that the graph-based approach can successfully handle composition tasks with multiple conditional constructs, meanwhile still generating solutions with a reasonably high quality.

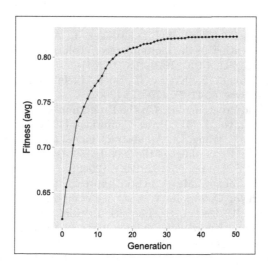

Fig. 2. Average fitness per generation with artificial optimum.

5 Conclusions

This work has discussed an Evolutionary Computing approach to Web service composition with conditional constraints that represents solution candidates as Directed Acyclic Graphs, as opposed to the previously explored tree-based representation. The increased flexibility of the graph representation requires specialised versions of mutation and crossover operators to in order to maintain the functional correctness of the solutions, and algorithms for accomplishing this were proposed. The graph-based approach was compared to an existing tree-based composition method, with results showing that the graph-based approach executes significantly faster than the tree-based approach for all datasets, even reaching a difference of two orders of magnitude for the largest dataset. The resulting solution qualities for both approaches, on the other hand, were found to be generally slightly higher when using the tree-based approach. Another experiment that tests the ability of the graph-based approach to handle composition tasks with multiple conditional constraints was also conducted, at the same time analysing the convergence behaviour of the approach. This test has shown that the graph-based approach can successfully produce solutions of a reasonably high quality for the more complex composition request, even though it could not reach the known global optimum. Consequently, future work in this area should explore alternative designs for the genetic operators used in the evolution process, likely resulting in improved fitness levels for the solutions produced. Sensitivity tests for parameters should also be conducted in the future.

References

1. Brown, N., McKay, B., Gilardoni, F., Gasteiger, J.: A graph-based genetic algorithm and its application to the multiobjective evolution of median molecules. J. Chem. Inf. Comput. Sci. **44**(3), 1079–1087 (2004)
2. Gottschalk, K., Graham, S., Kreger, H., Snell, J.: Introduction to Web services architecture. IBM Syst. J. **41**(2), 170–177 (2002)
3. Koza, J.R.: Genetic Programming: On the Programming of Computers by Means of Natural Selection, vol. 1. MIT Press, Cambridge (1992)
4. Ma, H., Schewe, K.D., Wang, Q.: An abstract model for service provision, search and composition. In: Proceedings of the 2009 IEEE Asia-Pacific Services Computing Conference (APSCC), pp. 95–102. IEEE (2009)
5. Ma, H., Wang, A., Zhang, M.: A hybrid approach using genetic programming and greedy search for QoS-aware Web service composition. In: Hameurlain, A., Küng, J., Wagner, R., Decker, H., Lhotska, L., Link, S. (eds.) Transactions on Large-Scale Data- and Knowledge-Centered Systems XVIII. LNCS, vol. 8980, pp. 180–205. Springer, Heidelberg (2015). doi:10.1007/978-3-662-46485-4_7
6. Mabu, S., Hirasawa, K., Hu, J.: A graph-based evolutionary algorithm: genetic network programming (GNP) and its extension using reinforcement learning. Evol. Comput. **15**(3), 369–398 (2007)
7. Menasce, D.: QoS issues in Web services. IEEE Internet Comput. **6**(6), 72–75 (2002)

8. Milanovic, N., Malek, M.: Current solutions for Web service composition. IEEE Internet Comput. **8**(6), 51–59 (2004)
9. Nicolaou, C.A., Apostolakis, J., Pattichis, C.S.: De novo drug design using multi-objective evolutionary graphs. J. Chem. Inf. Model. **49**(2), 295–307 (2009)
10. Perrey, R., Lycett, M.: Service-oriented architecture. In: 2003 Proceedings of the Symposium on Applications and the Internet Workshops, pp. 116–119. IEEE (2003)
11. Rodriguez-Mier, P., Mucientes, M., Lama, M., Couto, M.I.: Composition of Web services through genetic programming. Evol. Intell. **3**(3–4), 171–186 (2010)
12. Silva, A.S., Ma, H., Zhang, M.: GraphEvol: a graph evolution technique for Web service composition. In: Chen, Q., Hameurlain, A., Toumani, F., Wagner, R., Decker, H. (eds.) DEXA 2015. LNCS, vol. 9262, pp. 134–142. Springer, Heidelberg (2015). doi:10.1007/978-3-319-22852-5_12
13. da Silva, A.S., Ma, H., Zhang, M.: A GP approach to QoS-aware Web service composition including conditional constraints. In: 2015 IEEE Congress on Evolutionary Computation (CEC), pp. 2113–2120. IEEE (2015)
14. Sohrabi, S., Prokoshyna, N., McIlraith, S.A.: Web service composition via the customization of golog programs with user preferences. In: Borgida, A.T., Chaudhri, V.K., Giorgini, P., Yu, E.S. (eds.) Conceptual Modeling: Foundations and Applications. LNCS, vol. 5600, pp. 319–334. Springer, Heidelberg (2009). doi:10.1007/978-3-642-02463-4_17
15. Wang, L., Shen, J., Yong, J.: A survey on bio-inspired algorithms for Web service composition. In: IEEE 16th International Conference on Computer Supported Cooperative Work in Design (CSCWD), pp. 569–574. IEEE (2012)
16. Wang, P., Ding, Z., Jiang, C., Zhou, M.: Automated Web service composition supporting conditional branch structures. Enterp. Inf. Syst. **8**(1), 121–146 (2014)
17. Wohed, P., Aalst, W.M.P., Dumas, M., Hofstede, A.H.M.: Analysis of Web services composition languages: the case of BPEL4WS. In: Song, I.-Y., Liddle, S.W., Ling, T.-W., Scheuermann, P. (eds.) ER 2003. LNCS, vol. 2813, pp. 200–215. Springer, Heidelberg (2003). doi:10.1007/978-3-540-39648-2_18

Recommendation in Interactive Web Services Composition: A State-of-the-Art Survey

Meriem Kasmi[1(✉)], Yassine Jamoussi[1,2], and Henda Hajjami Ben Ghézala[1]

[1] Riadi Lab, ENSI, University of Manouba, Manouba, Tunisia
{meriem.kasmi,yassine.jamoussi,henda.benghezala}@ensi.rnu.tn
[2] Department of Computer Science, College of Science, Sultan Qaboos University,
PO Box 36, Al-Khoudh 123, Muscat, Oman
yessine@squ.edu.com

Abstract. With the increasing adoption of Web services, designing novel approaches for recommending relevant Web services has become of paramount importance especially to support many practical applications such as Web services composition. In this paper, a survey aiming at encompassing the state-of-the-art of interactive Web services composition recommendation approaches is presented. Both Web services composition and recommender systems concepts are introduced and their particular challenges are also discussed. Moreover, the need of using recommendation techniques to support Web services composition is also highlighted. The most relevant approaches dedicated to address this need are presented, categorized and compared.

Keywords: Web services composition · Interactivity · Recommender systems

1 Introduction and Motivation

Nowadays, Internet has globally proven to be a powerful platform where users can find all information they need. In fact, anyone can have access to the World Wide Web, and people everywhere are expressing their ideas and opinions through Internet. Even companies do not escape this rule as they can encapsulate their business processes and publish them as services using a Web service format [1]. This technology has become a de facto way for sharing data and software as well as integrating heterogeneous applications. Consequently, the number of Web services is tremendously increasing. According to the Web services search engine Seekda[1], there are 28,606 Web services on the Web, offered by 7,739 different providers as of August 2, 2011. Furthermore, several Web services publication websites have appeared such as WebServiceList[2] and XMethods[3]. This large number of Web services available has led to a challenging problem. That

---

[1] http://webservices.seekda.com/.
[2] http://www.webservicelist.com.
[3] http://www.xmethods.net/.

© Springer International Publishing AG 2017
D. Bridge and H. Stuckenschmidt (Eds.): EC-Web 2016, LNBIP 278, pp. 170–182, 2017.
DOI: 10.1007/978-3-319-53676-7_13

is, users have to choose the best Web service satisfying their needs and this is not easy due to this choice explosion. Moreover, due to the complexity and the diversity of users' demands, a single Web service is usually unable to respond to a specific user request. Thus, one interesting feature is the possibility to create new value-added Web services by composing other existing ones. This process called Web services composition (WSC) has become of paramount importance in the domain of Web services. It aims at integrating fine-grained Web services into large-grained composite ones. Currently, WSC has been heavily studied from both industrial and academic fields. This research area is drawing more and more attention in order to obtain the most relevant WSC and this is the rationale behind this paper.

Our paper focuses on the recommendation in the interactive WSC and offers a survey of state-of-the-art of recommendation approaches to support interactive WSCs. Then, it provides a classification of current approaches of interactive WSC recommendation, through which, we hope to contribute in the future research in the area of interactive WSC recommendation.

The remainder of this paper is organized as follows. Section 2 overviews some background information on WSC and recommender systems concepts and describes WSC recommendation in general. A classification of the interactive WSC recommendation approaches is presented in Sect. 3. Section 4 reports the comparative evaluation of the mentioned approaches. Section 5 gives a discussion of the evaluation and finally Sect. 6 sums up the conclusion.

2 Preliminary Concepts

The aim of this section is to give an outline of the key concepts and terminologies that will be needed in the rest of this paper. This will form a basis for the later sections.

2.1 Web Services Composition

According to the W3C[4] (World Wide Web Consortium), a Web service is "a software system identified by a Universal Resource Identifier (URI), whose public interfaces and bindings are defined and described using XML. Its definition can be discovered by other software systems. These systems may then interact with the Web service in a manner prescribed by its definition, using XML based messages conveyed by Internet protocols". To meet users' requirements, Web services can then be composed as new value-added and cross-organizational distributed applications [2].

The WSC process consists of four steps: planning, discovery, selection, and execution [3]. Planning determines the execution order of tasks. Discovery finds the candidate services for each task. Selection selects the best services from the discovered ones and finally the plan is executed.

---

[4] https://www.w3.org/.

WSC can be performed manually, automatically, or semi-automatically. For the first approach, where Web services are entirely composed by hand, users need to be technically skilled. This kind of composition is time-consuming and error-prone without any guarantee that the result will really satisfy the user's demands [4]. In contrast, in automated composition, the whole process is automatically performed without any user intervention required. However, realizing a fully automated WSC presents several open issues. It faces the indecision problem when, the system finds two or more functionally-similar candidate Web services. In this case, the system cannot make a choice, and therefore it needs to involve users [5]. The last approach aims at assisting users in the composition procedure. This composition being halfway between the previous two types is called also interactive WSC. Interactive WSC comes therefore to resolve the situation by addressing particular issues, for instance the difficulty of selecting a relevant service among the many available ones.

To sum up, manual composition seems to be, at first glance, the most adaptable to users' needs because it offers them the possibility to define everything as they want. However, it requires a good level of programming knowledge. In this situation, users who are tech-novice must be rejected. With the emergence of the Web 2.0, the composition process has become much more end-user oriented. In fact, this wave of Web has brought new technologies for end-users using graphic tools such as mashups. This technology has emerged as a promising way to enable end-users to combine easily services, in short time and obtain scalable results. Due to these advantages, mashups have become prevalent nowadays and therefore a number of mashup repositories have been established namelyProgrammableWeb.com, myExperiment.org, and Biocatalogue.org. In these repositories, a large number of published services are offered. For example, to April 20 2016, the largest Web services repository ProgrammableWeb.com possesses 7.806 Mashups. Moreover, several commercial mashups development environments were developed such as Yahoo pipes[5] and IBM Mashup Center[6].

2.2 Recommender Systems

The basic idea of recommender systems is to provide users with the most relevant items. They have become a rich research area since the 1990s when the first recommender system, Tapestry, was developed. Recommender systems have also re-shaped the world of e-commerce. In fact, many e-commerce sites such as Amazon and eBay are using recommender systems to suggest products to their customers and guide them in their choice of products to purchase.

To develop new recommendation approaches, much work has been done in both industry and academia. Classic recommendation approaches are usually classified into three categories: Content-based recommendation in which the user is recommended items similar to the ones he/she liked before, collaborative filtering-based recommendation in which the user is recommended items that

---

[5] http://pipes.yahoo.com/.
[6] https://www.ibm.com/support/knowledgecenter/SSWP9P/welcome/.

like-minded users preferred in the past and hybrid approaches combine collaborative and content-based methods. This last category helps to avoid shortcomings of both content-based and collaborative approaches and incorporates the advantages of these two methods.

Over the past few years, both Web services and recommender systems have been active research areas. The marriage between those two concepts has led to the application of recommender systems to Web services. Thus, we talk now about Web service recommender systems which are very useful especially that the available Web services search engines has poor recommendation performance. In fact, those search engines ignore non-functional characteristics of Web services [6] and using them, users should enter correct queries because they are keyword-based. Investing in the Web services field, current recommender systems focus mainly on Web services discovery and selection. We witness a wide range of papers about Web service recommendation mainly Web service discovery [7,8] and selection [9,10]. But, it is worth noting that there is not relatively large number of researches devoted to recommender systems for interactive WSC. Such a recommender system may be highly useful to the community especially that since the dawn of Web 2.0, end-users are more and more involved in the composition process in an interactive manner.

In this work, we are particularly interested in recommender systems for interactive WSC. We give thus a classification of their approaches in the following section.

3   Classification of Interactive WSC Recommendation Approaches

This section summarizes related studies that use recommendation to enhance interactive WSC. The classification of these approaches is mainly based on the new emerging trends in recommender systems dedicated to interactive WSC. That is why we have not mentioned classic approaches above. Categories of interactive WSC recommendation approaches presented are: context-aware, social network-based, quality-based, category-aware, time-aware and process-oriented interactive WSC recommendation approaches.

3.1   Context-Aware Interactive WSC Recommendation Approach

Context refers to any information that can be used to characterize the situation of entities (users or items). Examples of such information are location and time. Contexts can be explicitly provided by the users themselves or implicitly inferring by the system. Considering contextual information can be very useful and, in certain circumstances, not using this information in the recommendation process can disorient the recommendation results. For example, let's consider a travel recommender system. When asked for a vacation recommendation, it can give illogical suggestions if it ignores the temporal context of the request. The vacation recommendation in the winter can significantly differ from the one in the summer, so it is crucial to take into account this contextual information. This is the mission of context-aware recommender systems (CARS) which generate more relevant WSC recommendations by incorporating contextual information of the user.

3.1.1   Zhao(2010)

Zhao et al. [11] provided a comprehensive and extensible platform for service consuming and navigation named HyperService. With HyperService, non-technical users can easily search and navigate among services as they surf a document web with hyperlinks. Based on the input keywords and their navigation context, a set of relevant services is recommended. Every time users select a service among the suggested set, another set of services is recommended in accordance with the selected service. Thus, interactively-assembled Web services are brought to end users through a web2.0 style interactive user interface. Semantic Engine is the function kernel of HyperService. It provides the functions of automatic relation discovery, user behavior analysis (usage count, the rating score, etc.), service accessibility by keyword searching as well as context aware recommendations performed thanks to content-based methods. Services that fit user's current navigation context, having the best global importance factors (the more times a service is linked to other services, the more popular it is) and are of users' inferred interests are recommended and displayed to users in an interesting and user-friendly way.

3.2   Social Network-Based Interactive WSC Recommendation Approach

A social network is a graph representation of interactions between entities. This network has the potential to assume an important role in helping in decision-making situations. Let us recall some decisions that we make in our daily life, such as buying a new product or applying for a job in a particular company. Intuitively, we often ask our friends who have already had experience with that product or that company for their opinions. Therefore, social networks influence and even can change our views in decision-making scenarios. This awareness was fostered in numerous academic fields such as recommender systems. The idea of using social networks in recommender systems stems from other additional advantages mentioned in [17]. We summarize in what follows such prominent works.

3.2.1   Maaradji (2010)

Maaradji et al. [5] introduced the SoCo framework which relies on the retrieved knowledge from social networks modeling users' interactions to advise end-users on which Web services to select as they interactively undertake WSC tasks. Through a GUI offering drag/drop functionality for mashups, end-users receive a sorted list of candidate services that are relevant to be successor of the current one. They end up with a composition diagram representing the final WSC.

SoCo provides dynamic recommendations through an integrated social-aware recommender system. Its recommendations are built upon user profile (containing his/her interests, preferences, and the history of his interactions with the system) and social proximity extracted from social networks as well as the previously-built compositions. This information is used to estimate a recommendation confidence. Web services recommended are the most trusted ones i.e. having high recommendation confidence values.

3.2.2 Xu (2013)

Xu et al. [12] leveraged multi-dimensional social relationships among users, topics, mashups, and services to recommend the services needed to build mashups. Using a coupled factorization algorithm, services needed for the current mashup construction are predicted, ranked and delivered to the end-user at once. The recommended services are not functionally similar. They are rather the whole services needed and they are delivered to end-users not step-by-step but at once based on users' functional specifications and implicit requirements which are inferred by the topic model. Users have just to select proper services and compose them by performing "drag and drop" actions.

3.3 Quality-Based Interactive WSC Recommendation Approach

So far, interactive WSC recommendation approaches have not focused on the internal properties. Precisely, quality issues have not been invoked and there is a lack of proposals addressing quality-based interactive WSC recommendation. However, quality assessment may be instrumental when selecting services for composition. For instance, in case two services are functionally-similar, quality can be a discriminating factor. In line with this view, dealing with quality issues in the interactive WSC recommendation may be a promising research area.

3.3.1 Picozzi (2010)

Picozzi et al. [13] proposed a model aiming at supporting end-users in the recognition of the most suitable components. The quality-based proposed recommendation approach computes the quality of mashups to produce high-quality mashup recommendation. This value is actually an aggregated quality measure calculated on the basis of the quality of each component in the mashup. End-users who have already shaped final or intermediate mashup can get mashups recommendations about possible extensions of a given mashup. They can extend a particular mashup based on a certain recommendation and continue to extend the obtained mashup by considering other recommendations, realizing thus an interactive composition. The recommended mashups are ranked on the basis of a quality-driven recommender algorithm.

3.3.2 Cappiello (2012)

Cappiello et al. [14] illustrated the incorporation of quality based recommendations in the mashup development process to enable end-users to complete and/or improve their mashups. Thus, an assisted composition process in which quality and role of the candidate services are the driver of mashup recommendations was stressed in this work. As in [13], the quality of the composition is computed as a weighted aggregation of the quality of the single components. Weights reflect roles i.e. importance of each candidate service within the composition. Once the user selects the first candidate component, the quality-based ranking algorithm is executed. It proceeds according to two steps: (i) the categorization of the component to include in the current mashup using collaborative filtering mechanisms (ii) the selection of a particular component, belonging to the defined category in (i). Another interesting functionality of this algorithm lies in recommending similar but higher-quality compositions when applied on final ready-to-use mashups.

3.4   Category-Aware Interactive WSC Recommendation Approach

A WSC is a mixture of functionally-different Web services. It is quite obvious thus that the recommendation result contains services from various categories. For example, we find mapping services such as the Google maps service and auction services such as eBay.

However, most existing interactive WSC recommendation approaches do not provide candidate services ranked per category; they are given all in a single diverse list. This can lead to meaningless service ranking. Additionally, mashup composers are usually not clear about which categories they need. As long as relevant service categories are not explicitly provided, the user friendliness of recommendation will be decreased [15].

3.4.1   Xia (2015)

A novel category-aware service recommending method is proposed in [15]. It is actually a three-step approach to overcome the aforementioned restrictions and offers a performing category-aware service recommendation for mashup creation. In fact, after receiving a requirement text from a user, the category-aware recommendation engine analyzes it to infer the categories of services that are going to be implied in the mashup composition task. Then, the engine searches for candidate services belonging to the categories that it has inferred, and ranks them within these categories. Finally, the recommendation engine returns "per category service candidate ranking lists". The user selects from each category a service and thus the composition is executed. Once a mashup requirement is received, the service category relevance prediction starts. Combining machine learning and collaborative filtering, the approach decomposes mashup requirements and explicitly predicts relevant service categories. This implies that the problem where users are not clear about the needed service categories for mashup creation is hence solved through this approach. Finally, based on a distributed machine learning framework, services are recommended in the form of "Category per candidate service ranking lists". Hence, the meaningless service ranking issue is overcome.

3.5   Time-Aware Interactive WSC Recommendation Approach

Popular Web service ecosystems such as ProgrammableWeb.com are extremely dynamic and continuously evolving over time. This is due to the large number of incoming services joining the repository, simultaneously with many others perishing, becoming thus deprecated. For example, as we have mentioned before, to April 20 2016, there are 7.806 mashups available in ProgrammbeWeb.com but 1.530 of them are deprecated. This situation has led to the emergence of few but valuable efforts centered around time-aware recommendation approaches.

3.5.1   Zhong (2015)

Zhong et al. [16] extended their work in [15] to include the time factor reflecting the scalability of the ecosystem. Based on their model in [15], they developed a time-aware service recommendation framework for mashup creation. It is composed of three components: temporal information (TI) extraction, mashup-description-based collaborative filtering (MDCF) and service-description-based content matching (SDCM).

TI predicts service activity in the near future based on usage history. The predicted value corresponds to the service popularity score in recent time frame. To do this, TI predicts topic i.e. category activity first and then infers the service activity because directly predicting service activity will face the sparseness problem.

MDCF recommends services using collaborative filtering techniques applied on mashups having similar functional descriptions with the functional requirements of the new required mashup. Once the set of most similar historical mashups is obtained, the relevance score of services with respect to the new required mashup can be evaluated.

As for SDCM, it computes content similarity between the functional requirements of the new required mashup and the content description of services based on LDA (Latent Dirichlet allocation).

Popularity scores from TI and relevance scores from MCDF and SDCM are integrated to generate the ranked recommended list of services for the new required mashup.

3.6   Process-Oriented Interactive WSC Recommendation Approach

In some cases, users know which services should be selected but their arrangement in a correct workflow can be a tricky task. These approaches aim at supporting users in these situations.

3.6.1   Wijesiriwarna (2012)

Wijesiriwarna et al. [18] proposed a guided process-oriented mashup platform called SOFAP for software analysis composition. This platform allows different software projects stakeholders to access software analyses, compose them into workflows and execute the obtained composition. Every time users select a service, the recommendation engine provides him/her with the next possible services to add to the composition schema. In case of a wrong selection or following an incorrect control-flow pattern, recommendation engine gives a real-time feedback to the user. Once finished, the composed workflow is passed to the mashup run-time for the execution. Also, meaningful workflow templates are stored in the SOFAS repository, allowing future users to reuse the existing templates.

4   Comparative Evaluation

In order to evaluate the interactive WSC recommendation approaches classified in the previous section, the following criteria have been selected. Table 1 summarizes the result of the comparative evaluation.

The adopted criteria are described as follow:

- Personalization: It refers to the fact of recommending services according to users' interests. To do so, user behavior for example can be incorporated and analyzed into the recommendation model to generate more personalized and accurate services.
- Recommended items (Rec Items): Although the discussed approaches are all dedicated to generating relevant interactive WSCs, their outputs are not delivered in the same form. In [5,11,14], a ranked list of similar candidate services that are relevant to be successor of the current one are recommended. In [12], the recommended services are ranked but not functionally-similar. They are rather all the candidate services that are needed and are all delivered at once. In [18], similar services are recommended to add to the composition schema but the ranking issue was not invoked. In [13], ranked mashups are provided to end-users in order to help them in extending a particular mashup. In [15,16], ranked lists of candidate services organized into categories, are recommended. Thus, recommended items are either services (S) or mashups (M). They can be similar or different, ranked and categorized.
- Interactivity level: This criterion describes to which extent a user is involved in the interactive WSC. In [11], each time a user selects a service among the suggested set, another set of services is recommended in accordance with the selected service. Thus, the user is highly involved in the WSC process, since he/she should select candidate services one by one from the suggested sets to fulfill the WSC task. It is also the case in [5,14]. If the selection of candidate services is done simultaneously, interactivity level is lower. It is the case in [12,15,16]. When the user selects a whole mashup at one time, the interactivity is much lower such as in [13]. We respectively symbolize these three cases by (+++), (++) and (+).

Regarding the remaining criteria, they are those that we have explained above and adopted to categorize interactive WSC recommendation approaches. Those criteria are: Context awareness, Social Network awareness, Quality awareness, Category awareness, Time awareness and process orientation. In fact, we noticed that being of any class of approach does not mean excluding other classes. In contrast, this may improve recommendation accuracy and yield better results. Yet, there are different awareness levels towards these criteria. We distinguish 3 levels among the different studied papers: extreme awareness for those which particularly focus on that criterion, medium awareness for papers which adopt that criterion to refine more their recommendations but are not mainly structured around it and the no-awareness level for works which do not invoke that criterion at all. These three levels are respectively symbolized by: $(++)$, $(+)$ and $(-)$.

Table 1. Comparative evaluation of interactive Web services composition recommendation approaches

Criteria		Paper							
		Zhao et al. [11]	Maaradji et al. [5]	Xu et al. [12]	Picozzi et al. [13]	Cappiello et al. [14]	Xia et al. [15]	Zhong et al. [16]	Wijesiriwardana et al. [18]
Context awareness		++	-	-	-	-	-	-	-
Social network awareness		-	++	++	-	-	-	-	-
Quality awareness		-	-	-	++	++	-	-	-
Category awareness		-	-	+	-	+	++	+	-
Time awareness		-	-	-	-	-	-	++	-
Process orientation		-	-	-	-	-	-	-	++
Personalization		✓	✓	✓					
Rec. items	Component	S	S	S	M	S	S	S	S
	Categorized						✓	✓	
	Ranked	✓	✓	✓	✓	✓	✓	✓	
	Similar	✓	✓		✓	✓	✓	✓	✓
Interactivity level		+++	+++	++	+	+++	++	++	+++

5   Discussion

The work of Zhao et al. [11] presents the main advantage of taking into account context in their recommendation approach. This feature is crucial and can even leverage recommendation process because users' interests can change if ever being in a particular place or at a particular date. Recommendations are also personalized in this work. Nevertheless, its composition process requires high user involvement.

As for [5,12], they are both social-based but Xu et al. [12] exploited also the idea of category awareness to have much better results. Besides, they both make use of social information to support service recommendation but there is a difference pertaining to their models of relationships. In [12], relationships are multi-dimensional including users, topics, mashups, and services while, in Maaradji's [5], the social network models only users' interactions. Unlike [5], the recommended services are not functionally similar in [12]. They are all the services needed for the composition, delivered to end-users at once. This is an interesting feature offered by Xu's work keeping an effective level of interactivity in the WSC task. As a future outlook, Xu et al. announced their intention to build an online collaboration platform based on their approach. According to them, recommendations will be improved since the model will be more used, and thus it will collect more useful information on composition patterns. This is an appealing idea to which we are also interested.

Picozzi' work [13] and Cappiello's work [14] are among the few proposals addressing quality issues in interactive WSC recommendation. They both proposed extensions of users' mashups in order to enhance the overall mashup quality. This latter is perceived as a weighted aggregation of single components qualities. Weights reflect roles i.e. importance of each candidate service within the composition. Thus, these two works present the advantage of, in addition to being quality based, they are also role-based. In [13], the interactivity level is very low because its recommendations are high-quality ranked mashups to add to the current one. In contrast, users are highly involved in WSC in [14] since they have to select for each service belonging to their current mashup, another one from a ranked list of similar services. We note also that [14] performs an automatic categorization to recommend users with more relevant services. However, neither [13] nor [14] have involved personalization in their recommendation method.

Xia et al. [15] proposed a category-aware service recommending method. Its leading advantages are the fact that common restrictions (meaningless service ranking and low user friendless of recommendations) within existing recommendation methods are alleviated in their approach. Services are recommended in the form of "Category per candidate service ranking lists", so end-users have just to select from each category the service they need and the composition will be executed. Experiments conducted by authors have proved that their approach not only improves recommendation precision but also the diversity of recommendation results. Zhong et al. [16] provide almost the same model but including the time dimension when recommending service for mashup creation. Precisely, recommendations are also "Category per candidate service ranking lists" having relevant service activity in the near future. Despite their good performance, these two approaches do not provide personalized recommendations.

Wijesiriwardana et al. [18] proposed a process-oriented interactive WSC recommendation approach. This work is different from the other presented ones in many aspects. We mention the collaborative dimension of the proposed platformas well as the composition model based on workflow. Moreover, the recommendations provided are dedicated to software projects members such as developers, architectures and testers. One attractive advantage here is that the recommendation engine gives a real time feedback to the user in case of a wrong selection or incorrect control-flow pattern. Thus, users are very well supported in this work.

To sum up, we conclude that works exploiting the multi-awareness concept are limited, and the few ones exploiting this aspect, invoked only category awareness as a supplementary aspect of awareness. As for collaboration, only Wijesiriwardana et al. proposed a collaborative platform to interactively perform the WSC task.

Xu et al. just announced their intention to build a collaborative platform based on their approach. Hence, there is a considerable room for improvement regarding recommendation approaches dedicated to interactive WSC. In fact, exploring multi-awareness and collaboration features can be a promising way to enhance recommendations and to build a new comprehensive recommendation approach that outperforms all existing ones.

6   Conclusion

This paper has provided a comparative survey about interactive WSC recommendation approaches. An introduction to recommender systems as well as to WSC was presented in which we particularly focused on semi-automatic WSC. We raised and highlighted the need for a synergy between recommender systems and interactive WSC. Therefore, we studied the most prominent emerging approaches of interactive WSC recommendation and classified them into categories. We tried to cover all interactive WSC recommendation categories to get the most exhaustive possible classification. We also supported this classification by several criteria in order to evaluate and compare the approaches. Finally, a summary of the comparison and evaluation of the approaches are presented and discussed. We concluded thereby that our survey can help researchers in identifying benefits and limitations of interactive WSC recommendation approaches in order to develop a new comprehensive approach that outperforms all existing ones. This novel approach can take profit of the collaboration to highly improve their recommendations.

References

1. Tang, J.: A service composabilty model to support dynamic cooperation of cross-enterprise services. In: Shen, W. (ed.) Information Technology For Balanced Manufacturing Systems. IFIP International Federation for Information Processing, vol. 220, pp. 213–222. Springer, Boston (2006)
2. Zou, G., Lu, Q., Chen, Y., Huang, R., Xu, Y., Xiang, Y.: QoS-aware dynamic composition of web services using numerical temporal planning. IEEE Trans. Serv. Comput. **7**(1), 18–31 (2014)
3. Claro, D.B., Albers, P., Hao, J.-K.: Web services composition. In: Cardoso, J., Sheth, A.P. (eds.) Semantic Web Services, Processes and Applications, pp. 195–225. Springer, New York (2006)
4. Sheng, Q.Z., Qiao, X., Vasilakos, A.V., Szabo, C., Bourne, S., Xu, X.: Web services composition: a decade's overview. J. Inf. Sci. **280**, 218–238 (2014)
5. Maaradji, A., Hacid, H., Daigremont, J., Crespi, N.: Towards a social network based approach for services composition. In: Proceedings of the IEEE International Conference on Communications, Cape Town, pp. 1–5 (2010)
6. Yao, L., Sheng, Q.Z., Ngu, A.H.H., Yu, J., Segev, A.: Unified collaborative and content-based web service recommendation. IEEE Trans. Serv. Comput. **8**(3), 453–466 (2015)

7. Ma, J., Sheng, Q.Z., Liao, K., Zhang, Y., Ngu, A.H.H.: WS-finder: a framework for similarity search of web services. In: Liu, C., Ludwig, H., Toumani, F., Yu, Q. (eds.) ICSOC 2012. LNCS, vol. 7636, pp. 313–327. Springer, Heidelberg (2012). doi:10.1007/978-3-642-34321-6_21

8. Xu, Z., Martin, P., Powley, W., Zulkernine, F.: Reputation-enhanced QoS-based web services discovery. In: Proceedings of the 2007 IEEE International Conference on Web Services (ICWS), Salt Lake City, UT, pp. 249–256 (2007)

9. Zheng, Z., Ma, H., Lyu, M.R., King, I.: WSRec: a collaborative filtering based web service recommender system. In: IEEE International Conference on Web Services, pp. 437–444 (2009)

10. Zheng, Z., Ma, H., Lyu, M.R., King, I.: QoS-aware web service recommendation by collaborative filtering. IEEE Trans. Serv. Comput. 4(2), 140–152 (2011)

11. Zhao, C., Ma, C., Zhang, J., Zhang, J., Yi, L., Mao, X.: HyperService: linking and exploring services on the web. In: Proceedings of the 2010 International Conference on Web Services, Miami, FL, pp. 17–24 (2010)

12. Xu, W., Cao, J., Hu, L., Wang, J., Li, M.: A social-aware service recommendation approach for mashup creation. In: Proceedings of the 20th IEEE International Conference on Web Services, Santa Clara, CA, pp. 107–114 (2013)

13. Picozzi, M., Rodolfi, M., Cappiello, C., Matera, M.: Quality-based recommendations for mashup composition. In: Daniel, F., Facca, F.M. (eds.) ICWE 2010. LNCS, vol. 6385, pp. 360–371. Springer, Heidelberg (2010). doi:10.1007/978-3-642-16985-4_32

14. Cappiello, C., Matera, M., Picozzi, M., Daniel, F., Fernandez, A.: Quality-aware mashup composition: issues, techniques and tools. In: Proceedings of the Eighth International Conference on the Quality of Information and Communications Technolog, Lisbon, pp. 10–19 (2012)

15. Xia, B., Fan, Y., Tan, W., Huang, K., Zhang, J., Wu, C.: Category-aware API clustering and distributed recommendation for automatic mashup creation. IEEE Trans. Serv. Comput. 8(5), 674–687 (2015)

16. Zhong, Y., Fan, Y., Huang, K., Tan, W., Zhang, J.: Time-aware service recommendation for mashup creation. IEEE Trans. Serv. Comput. 8(3), 356–368 (2015)

17. He, J., Chu, W.W.: A social network-based recommender system (SNRS). In: Memon, N., Xu, J.J., Hicks, D.L., Chen, H. (eds.) Data Mining for Social Network Data. Annals of Information Systems, vol. 12, pp. 44–74. Springer, New York (2010)

18. Wijesiriwardana, C., Ghezzi, G., Gall, H.: A guided mashup framework for rapid software analysis service composition. In: The 19th Asia-Pacific IEEE Conference on Software Engineering Conference, pp. 725–728 (2012)

Author Index

Bizer, Christian 73

Caielli, Andrea Luigi Edoardo 34
Casanova, Marco A. 87
Cobos, Carlos 129
Cornec, Matthieu 100
Cremonesi, Paolo 34

da Silva, Alexandre Sawczuk 154
Danini, Maxime 100
de Amorim, Franklin A. 87
Deldjoo, Yashar 34
Domingues, Marcos 3

Elahi, Mehdi 34

Ferrarons, Jaume 142

Gama, João 3
Goutorbe, Bruno 100
Grauer, Christelle 100
Gupta, Smrati 142

Hajjami Ben Ghézala, Henda 170
Hartmann, Sven 154
Horch, Andrea 61

Jakubowicz, Jeremie 100
Jamoussi, Yassine 170
Jannach, Dietmar 21
Jiao, Yang 100
Jorge, Alípio M. 3
Jugovac, Michael 21

Kasmi, Meriem 170

Larriba-Pey, Josep-Lluis 142
Lima, Rafael Franca 111
Lopes, Giseli Rabello 87

Ma, Hui 154
Matthews, Peter 142
Matuszyk, Pawel 3
Merchan, Luis 129
Meusel, Robert 73
Moghaddam, Farshad Bakhshandegan 34
Muntés-Mulero, Victor 142

Naveed, Sidra 21
Nunes, Bernardo Pereira 87

Ordoñez, Armando 129
Ordoñez, Hugo 129

Pereira, Adriano C.M. 111
Pernul, Günther 46
Petrovski, Petar 73
Primpeli, Anna 73

Richthammer, Christian 46
Romano, Sebastien 100

Soares, Carlos 3
Spiliopoulou, Myra 3

Vinagre, João 3

Weisbecker, Anette 61
Wohlfrom, Andreas 61

Zhang, Mengjie 154

Printed in the United States
By Bookmasters